TECHNICAL
RESCUE
OPERATIONS

Volume I: PLANNING, TRAINING, & COMMAND

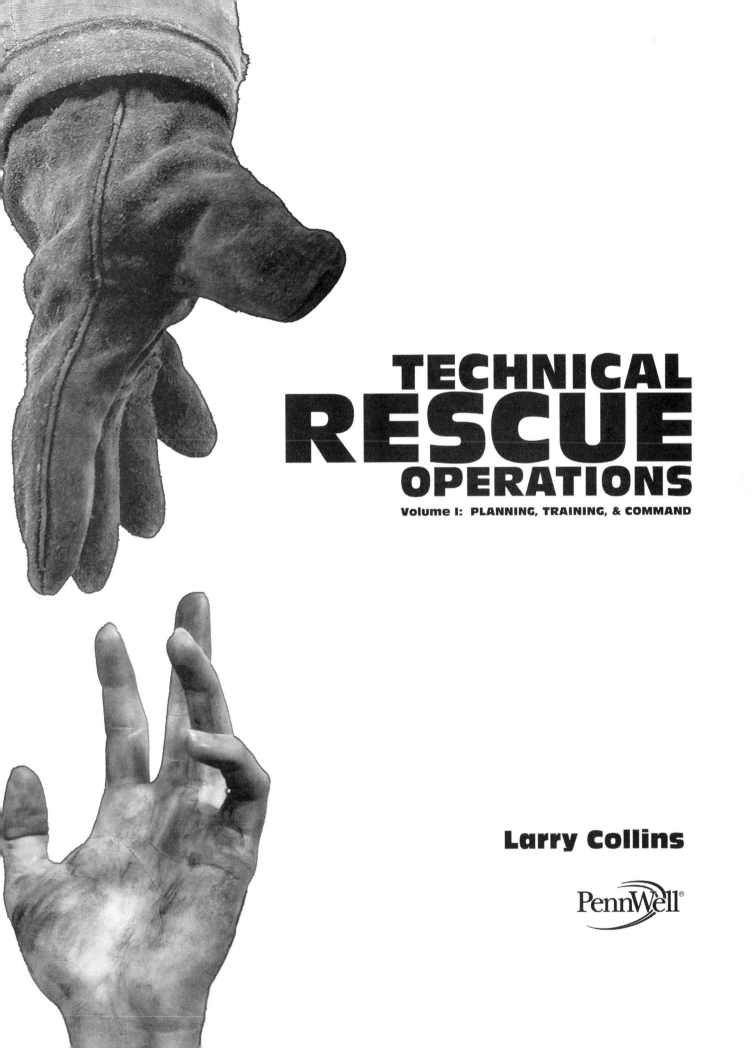

TECHNICAL RESCUE OPERATIONS

Volume I: PLANNING, TRAINING, & COMMAND

Larry Collins

PennWell®

Copyright© 2004 by
PennWell Corporation
1421 South Sheridan Road
Tulsa, Oklahoma 74112-6600 USA

800.752.9764
+1.918.831.9421
sales@pennwell.com
www.fireengineeringbooks.com
www.pennwell.com

Managing Editor: Jared d'orr Wicklund
Production Editor: Sue Rhodes Dodd
Cover Design: Ken Wood
Book Design: Wes Rowell

Library of Congress Cataloging-in-Publication Data

Collins, Larry, 1952–
  Technical rescue operations / by Larry Collins.
      p. cm.
  Includes bibliographical references and index.
  ISBN 1-59370-014-8 (8)
  1. Emergency management--United States. 2. Rescue work--United States--
Planning. 3. Search and rescue operations--United States. 4. Disaster relief--
United States. I. Title.

  HV551.3.C65 2004
  363.34'81'0973--dc22

                                                         2004011227

Printed in the United States of America

1 2 3 4 5 08 07 06 05 04

This book is dedicated to the innocent people who lost their lives in the events of September 11, 2001, and particularly those who gave their own lives in the service of others.

How can we adequately describe the heroics of the firefighters, police officers, and emergency medical service (EMS) personnel who looked up to see a nightmare unfolding before their eyes, and yet charged into that maelstrom of fire without hesitation? How can we fully characterize the actions of those who rushed in to save lives, even as bodies and debris rained down from the flaming towers? How can we properly depict the bravery of the leaders who held their ground in the danger zone to organize the emergency response to a *worst-case scenario* that would eventually kill thousands of citizens and hundreds of responders? And how do we begin to describe the selflessness demonstrated by those who escaped the first collapse only to lose their lives reentering the collapse zone to rescue trapped firefighters, police, EMS workers, and citizens?

To adequately describe these acts is impossible, no less difficult than describing the Grand Canyon to someone who never experienced sight. Perhaps Han-Jochen Blatte, chief of a German fire brigade, put it best when he wrote simply: "The firefighters from FDNY had only a small chance. But they used it completely for others and died themselves."

On a personal note, it should be mentioned that certain New York responders like Ray Downey, Dennis Mojuica, Thomas Langone, and other members of the Fire Department of New York (FDNY) and the New York Police Department (NYPD) greatly influenced this author's work before losing their lives on September 11, 2001.

Ray Downey holds a special place in this author's education and experience. As a co-instructor, co-author, and co-member of the original Federal Emergency Management Agency (FEMA) urban search and rescue

(USAR) response system, as a leader in the FEMA USAR National Response System, and as a mentor, Chief Downey greatly influenced me and many other fire/rescue personnel. Few people carried such a wealth of hard-wrought experience in the art and science of rescue. Fewer still had the motivation and the ability to communicate a vision for improving the future of the fire and rescue services (Fig. D–1).

Fig. D–1 FDNY Special Operations Command Chief Ray Downey Directing Operations at a Gas Explosion (Courtesy FDNY)

In his dual role as a task force leader of the New York FEMA USAR Task Force and operations chief of a FEMA USAR incident support team (IST), Downey would normally have played a lead role in the disastrous fire and collapse that resulted from the attack at the Pentagon on 9-11. But, as fate would have it, New York had already been struck and Downey was in the thick of things at the World Trade Center. He survived the collapse of the South Tower and was last seen with other FDNY members repeatedly entering the collapse zone to rescue citizens and downed firefighters, even though it was evident that the North Tower was also about to come down all around them.

One of Downey's final orders was an unequivocal and sharp-tongued demand for a younger officer to leave the collapse zone—now! Covered in dust and debris from the collapsed South Tower, Downey, a highly decorated officer just months away from retiring after 43 years of dedicated service to the FDNY and the United States, could have retreated from the collapse zone in time to avoid the collapse of the second tower.

But it was a zero-moment point; once Downey and the others committed themselves, return would not be possible. Downey and other FDNY members chose to stay in the fall zone to help others escape. They crossed the proverbial Rubicon and never came back.

Ray Downey vastly improved the nation's response to collapse disasters by helping develop the FEMA USAR task force system from concept to reality in just three years. He also wrote the gospel on collapse in his seminal book, *The Rescue Company*, and authored many informative articles about collapse rescue. It is no small irony that he himself died in the largest structural collapse in history, a disaster so devastating that it resulted in the response of thousands of firefighters and rescuers, including 25 FEMA USAR task forces from around the nation.

With Downey in those final moments were many members of FDNY rescue and squad companies, whose specialty included rescuing people from collapsed buildings, and who, under Downey's command, vastly expanded and improved their own capabilities and those of hundreds of other New York firefighters. It is another tragic irony that all five of the rescue companies in New York were lost with the other responders, and the FEMA USAR task forces that Ray Downey helped pioneer were assigned to aid in the search for all of them.

The valor demonstrated by the FDNY and the other emergency responders that day cannot be faked, learned, or bought. These people were the *real thing* (Fig. D–2).

Fig. D–2 The Crew of FDNY Rescue Company 3 Taking up after an All-hands Fire (Courtesy FDNY)

Contents
Volume I

Disclaimer

The training and activities related to *urban search and technical rescue* are potentially dangerous, frequently occurring in high-risk environments and conditions that could lead to serious injury and death if not properly organized and supervised. In the event of unforeseen circumstances such as earthquake aftershocks, secondary collapses, and the like, serious injury or death could result even *if* the activities of the reader are properly organized and supervised.

Therefore, the reader is expressly warned to consider and adopt all safety precautions that might be indicated by the activities herein and to avoid all potential hazards. During the course of training and emergency activities related to *urban search and technical rescue,* every individual involved should act as a safety officer and be constantly alert for any potential safety violations or problems.

The author, *Fire Engineering Books*, and those individuals and agencies who contributed information for this guide take no responsibility for the use or misuse of the material herein. Neither the writer nor the publisher warrant or guarantee any of the products described herein. The writer, contributors, and publisher take no responsibility for the loss, damage, injury, or death resulting from information contained or omitted from this guide. This guide should not be used as a manipulative training manual nor a substitute for professional instruction, manipulative training, and experience in realistic training environments. The manipulative methods depicted in this book should be conducted only by select personnel under the supervision of qualified instructors and team leaders.

By following any instructions contained herein, the reader willingly assumes all risks in connection with such instructions and activities. Although this book covers many aspects of urban search and technical rescue, it is not a basic SAR manual and should only be used as a guide. Although a number of official agencies are discussed and depicted in this guide, it is not a publication or training manual of any government or private agency. The author and the publisher shall not be liable for any special, consequential, or exemplary damages resulting, in whole or in part, from the reader's use of, or reliance upon, this material.

This book was written as a guide for qualified and experienced fire/rescue personnel and emergency managers, with no intention of contradicting the current policies of any agency or organization. The generic terms *he, him*, and *his* are used herein only for economy, recognizing that females are included in the ranks of fire/rescue agencies worldwide.

List of Acronyms

AHJ   (agency having jurisdiction)

BLS   (basic life support)

cfs   (cubic feet per second)

DROP   (Disaster Rescue Operations Plan)

ELT   (emergency locator transmitter)

EMS   (emergency medical service)

ESF   (Federal Emergency Support Function)

FCCF   (Fire Command-and-Control Facility)

FDNY   (Fire Department of New York)

FEMA   (Federal Emergency Management Agency)

FIRESCOPE   (Firefighting Resources of California Organized for Potential Emergencies)

FOG   (field operations guide)

GPS   (global positioning satellite)

HEED   (helicopter emergency escape device)

IAFF   (International Association of Firefighters)

IAP   (incident action plan)

IC   (incident commander)

ICS   (incident command system)

IDLH   (immediately dangerous to life and health)

IETRI   (International Emergency Technical Rescue Institute)

INSARAG   (International Search and Rescue Advisory Group)

IRB   (inflatable rescue boat)

IST   (incident support team)

KSA   (knowledge, skills, and abilities)

L.A.   (Los Angeles)

LACoFD   (Los Angeles County Fire Department)

LAFD   (Los Angeles Fire Department)

LAPD   (Los Angeles Police Department)

LAX   (Los Angeles International Airport)

LCES   (lookout, communication, escape route, and safety)

MCI   (multi-casualty incident)

MHFP	(multi-hazard functional plan)
MRE	(meals ready to eat)
NASAR	(National Association for Search and Rescue)
NFIRS	(National Fire Incident Reporting System)
NFPA	(National Fire Protection Association)
NOAA	(U.S. National Oceanic and Atmospheric Administration)
OCFA	(Orange County Fire Authority)
OES	(Office of Emergency Services)
OFDA	(Office of Foreign Disaster Assistance)
OSHA	(Occupational Safety and Health Administration)
PFD	(personal flotation devices)
PLS	(point last seen)
PPE	(personal protective equipment)
PWC	(personal watercraft)
RDM	(rapid decision-making)
RIC	(rapid intervention crew)
RPD	(recognition-primed decision)
SABA	(supplied air breathing apparatus)
SAR	(search and rescue)
SCBA	(self-contained breathing apparatus)
SCEC	(Southern California Earthquake Center)
SEMS	(Standardized Emergency Management System)
SFSAR	(swift-water/flood search and rescue)
SOG	(standard operating guideline)
SRC	(search and rescue coordinator)
STTU	(Specialized Tactical Training Unit)
TCO	(traffic collision over-the-side)
URM	(unreinforced masonry)
USAID	(U.S. Agency for International Development)
USAR	(urban search and rescue)
VATF-I	(Virginia Task Force I)
WMD	(weapons of mass destruction)

Preface

When *Technical Rescue Operations* was first proposed as a book series, one goal was to capture the essence of the dramatic sea-change in the philosophy about technical rescue that ultimately spawned the formalized development and consolidation of the disparate disciplines known as *rescue* and its younger cousin, USAR. It goes without saying that firefighters and rescuers in nations around the world today achieve unprecedented levels of capability, cooperation, information sharing, and consensus on rescue, firefighting, EMS, hazardous materials (hazmat), terrorism-consequence management, and incident command. In addition to the time-honored fire and rescue service values of hard work, dedication to duty, and willingness to sacrifice personal safety to help others, it can be said that the new levels of success are proportional to the new paradigm in urban search and technical rescue operations.

I have had the privilege and honor of participating in development, planning, instruction, and emergency response for technical rescues and disasters in Los Angeles (L.A.) County as a fire captain in the L.A. County Fire Department (LACoFD). I have worked across the United States as a participant in FEMA's national USAR system and on every continent as a member of a U.S. international USAR task force under the aegis of the U.S. Agency for International Development/Office of Foreign Disaster Assistance (USAID/OFDA). Based on these experiences, it's my observation that for all the differences in the world's fire and rescue services and the wide variety of emergency challenges they face, there are more similarities than differences among firefighters, rescuers, and their respective agencies everywhere. This is in part attributable to the new philosophy of rescue that has evolved during the past two decades.

The growing success rate for rescues achieved by fire and rescue agencies that have adopted proven strategies, tactics, and equipment cannot be disputed. It's clear that firefighters and rescuers whose agencies have adopted *best practices* (from wherever they may be found) are more likely to rescue victims alive and with reduced casualties among the rescuers themselves. This applies both to daily rescue emergencies and to major disasters involving missing and trapped victims.

With that in mind, the main concept of this four-volume series is to avoid producing strictly *nuts-and-bolts* instructional manuals. Many books covering major elements of rescue have already been written, so there is no need to reinvent that wheel.

With this in mind, Volume I is subtitled *Planning, Training, & Command* to reflect the preparatory steps necessary to make sure fire/rescue agencies have the infrastructure to support the development of urban search and technical rescue capabilities at different levels and strengths. Volume II (*Common Emergencies*) deals with a wide range of technical rescue operations faced on a daily basis by many fire/rescue agencies and also deals with major rescue operations that may require many hours to complete. These are not necessarily disasters *per se*, but they typically require a major commitment of trained resources to complete. Volume III covers *Disaster Rescue*, the kind of incidents that will be *once-in-a-career* events for most fire/rescue practitioners. But when they occur, the public expects local fire/rescue personnel to be prepared to manage them effectively all the same. Volume III concludes with the largest-scale rescue operations, those that occur during the course of disasters such as major earthquakes, floods, hurricanes, and tornadoes, etc.

The reader will find that these books are concerned with highlighting recent advances in strategies, tactics, and operational systems. They books review the use of equipment and apparatus as well as rescue operational planning, preparedness, and response. Selected case studies are presented that spotlight the growing recognition that urban search and technical rescue in its many forms plays a fundamental role in the world's fire and rescue services. These books are not intended as formal instructional manuals for high-risk manipulative training without qualified instructors. Manuals for technical operations such as emergency trench rescue, technical rope systems, dive rescue, and other related operations are already available from government agencies, standards-setting organizations, university presses, and private publishers, and I recommend that readers go to those sources when they need strictly manipulative training documents.

A prime goal of this series of books is to examine rescue lessons that have local, regional, national, and even international implications. As just one example, consider the development since 1990 of 28 FEMA USAR task forces, strategically located in major fire departments across the United States. This system provides incident commanders (ICs) with access to urban search and technical rescue resources that could only have been dreamed of in past decades. Recent man-made and natural disasters across the United States, combined with the fast pace of development and national readiness, have led some to say that the FEMA USAR task force system is among the most successful, cost-effective, and sustainable programs ever launched by the U.S. government. This is certainly one example of a collaborative effort between firefighters and rescuers from both coasts and the middle of the nation, who in this case came together to build a system that can now bring unprecedented levels of rescue expertise to each of this nation's fifty states within hours of a disaster.

On the international front, in recent years the U.S. government, working with three selected fire departments, has developed USAR task forces capable of deploying to other nations, operating under the aegis of the USAID and specifically under the direction of the OFDA. As of this writing, the internationally deployable USAR task forces are based in the Fairfax County (VA) Fire and Rescue Department, the Miami-Dade (FL) Fire and Rescue Department, and the County of Los Angeles Fire Department. These American teams (each of which is also a participant in the FEMA USAR task force system for domestic disasters) join USAR teams from dozens of nations around the world in an international system that responds to major disasters wherever they may occur.

To develop better standardization and coordination of these teams, the International Search and Rescue Advisory Group (INSARAG) was established in 1991. The group includes representatives from dozens of nations around the world organized into three regions: Europe and Africa, Asia-Pacific, and the Americas. Based on recommendations from INSARAG, in December 2002 more than 53 nations signed the United Nations General Assembly resolution on international aid, including USAR.

The continuing development of all these domestic and international disaster search and rescue (SAR) capabilities is unprecedented, and they are yet another example of the worldwide spread of the *rescue revolution*. The current level of domestic and international cooperation, information sharing, and operational effectiveness would have been unfathomable just two decades ago. Yet it is happening, and it continues to grow.

How the 9-11 Attacks Have Affected Rescue

The 9-11 terrorist attacks in New York City and at the Pentagon marked a turning point in history, not just for America but also for fire and rescue services around the world. These attacks and the major events that occurred in their wake have literally and figuratively changed the world in which each of us lives and works. Prior to September 11, 2001, the importance of local, regional, national, and international USAR systems was overlooked in some quarters. But in the aftermath of these terrorist attacks, it became clear that the growing emphasis on rescue was well-founded.

It's a measure of progress in urban search and technical rescue that local and regional fire/rescue resources, backed by 20 FEMA USAR task forces, were able to manage the consequences of the tremendous collapse rescue disasters that occurred in New York and Washington, without any loss of life among the firefighters and rescuers who responded in the hours and days following the collapse of the World Trade Center Towers and the Pentagon. This was true even after the deaths of so many firefighters and specialized rescue units of the FDNY and NYPD, including many of those assigned to the New York City FEMA USAR task force and some who were part of FEMA's USAR ISTs.

A New Paradigm in Rescue

The chronology of the so-called *rescue revolution* can be traced back for more than two decades. Beginning in the early 1980s, the fire service experienced the beginnings of a new paradigm in the management of rescue emergencies. It began with the recognition that fire departments in many regions have implied, or in some cases, legal responsibility to conduct rescue operations in a professional manner using sound tactics and strategies and the proper equipment. This responsibility should also ensure a reasonable level of safety for firefighters and other safety personnel. The revolution included the recognition that many traditional approaches to rescue were ineffective and fraught with unnecessary life hazards to the personnel called upon to conduct these operations.

The new rescue paradigm included a recognition that fire and rescue agencies have a responsibility to ensure timely rescue (e.g. rapid intervention) for firefighters and other safety personnel who themselves become lost, trapped, or injured during the course of fireground and rescue operations. The impetus for a new model of rescue was strengthened by repeated demonstrations that the application of formalized technical rescue programs improved survivability for trapped victims who were being rescued faster and with fewer preventable complications. In addition, rescuers armed with newfound knowledge about the conditions they faced and with effective solutions and good equipment were becoming trapped, lost, or injured less often. Finally, the new paradigm in rescue was actualized because the public, which has always relied on fire/rescue agencies to fix practically any emergency problem, has come to expect—and in many cases to demand—timely, professional, and effective rescue response from its local fire department and even from its state and federal governments.

As the new model for rescue spreads, many nations are experiencing a dramatic proliferation of fire department-based rescue and USAR programs. Some 20 years later, the typical fire department provides firefighters, rescuers, fire chiefs, and emergency managers with the tools to effectively manage most rescue emergencies. Specialized USAR or rescue units have the ability to handle highly complex SAR operations that once might have defied their efforts. During major disasters in many nations, the units are supported by a network of state and national teams trained and equipped to manage the most daunting rescue-related disasters. When national resources become overwhelmed and the nation experiences a disaster, it can request help and accept assistance from the international community standing ready to deploy dozens of USAR teams.

In short, the fire/rescue agencies of the United States and other modern nations have achieved the highest level of preparedness for rescue in history. And it's none too soon; these systems are being challenged by some of the most deadly acts of terrorism in the experience of mankind, even as more conventional rescue hazards like earthquakes, hurricanes, industrial explosions, dam failures, and other natural and man-made disasters continue to occur.

A win-win situation

This newfound focus on rescue has vastly improved the means by which firefighters and other rescuers in many nations are able to detect, locate, treat, and extract trapped victims in practically any situation. Mirroring this paradigm shift is an ever-expanding and correlative pattern of dramatic survival stories—successful rescues accomplished under conditions that might have resulted in failure before the advent of formal rescue programs. There's strong evidence connecting the two phenomena. At the same time that firefighters and rescuers are experiencing a revolution in the availability of better tools, training, and systems for locating and rescuing trapped people, more trapped people are surviving complex entrapment under conditions that once would have meant almost certain death.

Clearly there is a close link between today's highly advanced rescue capabilities and the higher survival rates for trapped victims. For those of us who have seen or experienced the success of complicated rescues that once would have defied us, no other explanation is necessary. These experiences leave us with an intrinsic understanding that a vigorous and pro-active approach to unusual rescue situations will result in the saving of lives that once would have been lost. We also understand that these new rescue capabilities might one day be called upon to save our own lives if things go wrong on the scene of an emergency and we ourselves become trapped. From the perspective of both the rescuers and the people we are sworn to serve, the development of better rescue capabilities everywhere is truly a *win-win* situation.

Figure P–1 shows a man trapped in an excavation collapse waiting to be rescued. Three hours later, the

man is removed from the excavation collapse with minor injuries. A miracle? or the result of good training, solid planning, and the development of an effective urban search and technical rescue system? (Fig. P–2)

Fig. P–1 Trapped Worker Awaits Extrication by USAR-trained Firefighters

Fig. P–2 Trapped Worker is Freed

Today, when a victim survives a particularly riveting entrapment, it's frequently characterized as *miraculous* by the news media and the public. But informed observers understand that the real miracle may be the sea-change in the way modern fire/rescue services (sometimes assisted by specialized state and federal teams) manage rescue operations.

The new approach to rescue emphasizes factors such as innovative training, better equipment, and more intelligent planning by well-informed decision-makers who constantly ask, "What if?" It's a philosophy that emphasizes adherence to accepted incident command system (ICS) and standard emergency management system (SEMS) principles. The new system increasingly relies on fully-staffed rescue companies and USAR units for daily fire/rescue operations and better adherence to personnel safety principles. These principles include the use of redundant safety systems, rapid intervention, and other related protocols.

In combination with the time-honored fire/rescue service traditions of rapid response, hard work, sacrifice, and the application of common sense, these new factors often lead to dramatic, well-planned rescues that look miraculous to the public. Although this book emphasizes skill, planning, good training, and experience as the building blocks for successful rescue operations, who among us can discount the occasional miracle—assuming one believes in such things—as a contributing factor?

The last two decades also saw the evolution of daring new programs such as state-sponsored swift-water rescue task forces, multi-agency swift-water rescue systems, and waterway rescue preplans. Improvements have been made in the ICS and new National Fire Protection Association (NFPA) technical rescue and firefighter health/safety standards have been developed. Another innovation is the growing use of fire/rescue helicopters for technical rescue. These are just a few of the many hallmarks of a dramatic new chapter in rescue.

Who has primary responsibility?

There were of course other goals for this text. Chief among them is to acknowledge the expanding role of modern fire departments as the primary providers of prompt and professional rescue in all its forms and in practically every environment in many regions of the nation. Today the task of locating and safely extracting victims from collapsed buildings, cliffs,

machinery, flood waters, burning buildings, and other daily technical SAR predicaments is managed successfully by fire departments across the nation.

When disaster strikes, local firefighters are usually the first to arrive and the last to leave, and they're expected to handle pretty much everything in between—often for extended periods without outside assistance. If protecting lives, the environment, and property (in that order) are the primary fire service priorities during rescue-related emergencies and disasters, it's self-evident that rescue is among the most important responsibilities of every firefighter. In Figure P–3, local firefighters and FEMA USAR task forces work together at the scene of the Oklahoma City bombing. Such coordinated efforts involving hundreds (or thousands) of firefighters and other rescuers, are a hallmark of the new emphasis on more effective disaster response.

Fig. P–3 FEMA USAR Task Forces in Oklahoma City

It's also a goal of this series of books to recognize the continued importance of mountain SAR teams, tunnel and mine rescue teams, industrial rescue brigades, dive rescue teams, and other groups called upon to perform rescue. Volunteers who dedicate significant amounts of their own time to the cause of improving local SAR capabilities staff many of these teams. In some places these teams remain the primary rescue resource. In many cases the best outcome for victims is ensured by partnerships between traditional SAR teams and municipal fire departments.

Regardless of locale, it's important for fire departments, rescue teams, industrial brigades, and other rescue resources to foster inter-agency cooperation and trust. It's important for them to agree on the use of unified command at rescue incidents for which they share responsibility. It's also critical that they agree on the principle that the only reason they exist is to serve the public, and therefore it's in everyone's best interest if they work together to ensure that the rescue of lost or trapped victims is *always* job number one.

This may be a bitter pill for some rescuers who aren't accustomed to working closely with other agencies, or for those who feel their team has been ordained as the only resource that will be allowed to do the job, even when other emergency resources are available. In some places, rescue (particularly wilderness rescue) has traditionally been a law enforcement function. That dynamic is changing in many regions as local fire departments gain expertise in managing the detection and rescue of lost or trapped victims in wilderness and mountain rescue (Fig. P–4).

Fig. P–4 Modern Firefighters Manage a Wide Range of Mountain Rescue Operations

Rescue Lessons and a Historical Perspective

Yet another goal of *Technical Rescue Operations* is to convey lessons learned through years of experience by fire and rescue agencies with active USAR and technical SAR programs. Particular attention is paid to the lessons learned by members of busy rescue companies, USAR task forces, swift-water rescue teams, helicopter rescue units, collapse teams, confined-space teams, mine and tunnel teams, mountain rescue teams, and others whose responsibilities include urban search and technical rescue on a regular basis.

Sometimes a review of history is helpful in understanding and achieving the full measure of that which is possible in the future. There is a natural tendency for humans to assume that the present condition is the way it's always been, even in fields of work or study where rapid change should be evident to the dispassionate observer. In the pursuit of progress, it is therefore helpful to disrupt our tendency toward inertia by shaking up the system by disproving antiquated assumptions. One way to do this is to contrast the past with the present and then look forward with a renewed expectation that more drastic changes are still coming. We should be prepared to shape those changes to our advantage. Therefore, another purpose of this series is to provide a historic perspective on rescue and disaster response.

Physical entrapment is a condition known to people throughout the ages and not just in relation to slavery or imprisonment. One can only imagine the variety of ways in which people managed to get themselves trapped, pinned, or otherwise stuck in daily life during ancient times. Many ancient entrapment scenarios undoubtedly mirrored the kind of rescues that confront modern fire/rescue personnel, and some will continue to occur in the future. This is why it's important to pass on lessons learned to the next generations of rescuers.

Disasters have plagued man since he first walked the earth, and they continue to affect the way we live practically every day. From earthquakes that leveled ancient cities like Jericho, Troy, and Lisbon to the lost city of Ubar, which literally dropped below the desert sands into a giant sinkhole possibly during an earthquake, disasters have affected many lives. From great floods that may have inspired Biblical stories to volcanic eruptions that erased highly evolved civilizations, disasters have managed to change the course of human history. Many experts agree that disasters affect mankind with ever-greater frequency. This in part is related to the growing human population. There is also a correlation with our seeming insistence on living in the most disaster-prone locations.

In addition to the direct benefit to victims, effective disaster rescue operations carry a great positive societal impact. Students of government, politics, history, disaster response, and terrorism understand that one of the great questions when disaster strikes is: Will the government's response be effective, or will it be chaotic and haphazard? Around the world, governments have fallen as a result of ineffectively dealing with disasters.

A government that is unable to cope with disaster is vulnerable to public opinion that naturally questions the ruler's ability to properly rule. That is a topic that governments generally wish to avoid, and the best way to avoid the question is through effective disaster response. If local, state, and national governments are observed responding to disasters quickly, effectively, and through the exercise of effective management, the populace naturally develops a sense of trust. If, on the other hand, the government's response to disaster is horribly botched, the public will be less tolerant and may even revolt. This perhaps is a simplistic explanation of the pitfalls of poor response to disasters, but it's no less true.

This concept is especially true with regard to terrorism. One of the aims of terrorists is to destabilize a situation and create doubt about the government's ability to protect the public's safety. Terrorists can

sometimes gain a foothold if they can somehow demonstrate that a government is unable (or unwilling) to take appropriate steps to ensure public safety. Obviously the best solution is to thwart terrorists before they can act, a step known as *crisis management* in law enforcement parlance.

In cases where terrorists succeed in attacking, the next most important solution is a rapid, measured, appropriate, and effective emergency response (known as *consequence management*) to ensure that trapped, injured, and missing people are given the best care available in the shortest amount of time. In the event of an actual terrorist attack, a rapid and effective response to the consequences is the best way to defeat the aims of the terrorists. Once again, a study of past history is helpful in understanding the importance of a proper response. As time marches forward, there are ever-increasing risks for human entrapment; yet there are also ever-expanding options to properly manage these emergencies to stack the odds in the victim's favor. That is why this book includes reflections of historical disasters.

Modern perspectives

Another goal of this book series is to give a nod to forward-looking individuals and agencies that recognized the need to establish dedicated formal rescue systems as early as the start of the 20th century. They recognized the advantage of units and teams specializing in various forms of SAR, particularly the skills and equipment needed to quickly locate and remove firefighters and other rescuers who became lost or trapped during the course of emergency operations.

The proud traditions begun by those early pioneers continue today, though it's safe to assume that few would have anticipated the variety of forms rescue has assumed in the intervening decades. Nor is it likely that they would have predicted the expanding role of fire departments as the primary provider of all-risk SAR service in many parts of the world. For example, swiftwater and flood rescue are now considered a normal part of the fire department's mission (Fig. P–5).

Fig. P-5 Crew of Swiftwater Rescue Team Search for Victims Missing in Flood

Who would have predicted that swift-water rescue and river/flood rescue would become primary fire department missions in arid and semi-arid places like Phoenix, Las Vegas, Albuquerque, and southern California? Swift-water rescue capabilities have been formalized nationwide to a surprising extent. Two decades ago, swiftwater rescue was practically unknown to most fire departments and many rescue teams. In fact, many scoffed at the idea that firefighters would one day be equipped and trained to rescue people from raging flood waters. Today, many fire engines and ladder companies are equipped with personal floatation devices, floating rope, inflatable rescue boats, and other tools of the trade for swift-water rescue. Many laughed at the concept of waterway rescue preplans to increase the speed and effectiveness by which modern fire/rescue agencies would deploy within minutes. Agencies now have preplanned strategic rescue points to intercept victims being swept away in streams, rivers, flood-control channels, and aqueducts; today all these things are commonplace.

Who would have anticipated the day when municipal fire departments provide expert management of cliff rescues and other technical wilderness rescues in mountainous regions? Who would have predicted the extent to which modern fire departments conduct rescues along the coast and the frequency with which they assist the U.S. Coast Guard during disasters at sea?

How many firefighters signed onto municipal fire departments expecting to be assigned to units that extract victims from deep within abandoned gold mines, submerged tunnels, collapsed freeways, or downed airplanes in rugged mountains? Who would have foreseen the extent to which modern fire departments provide expert management of the most complex rescues that occur on a daily basis across the United States and other nations? Figure P–6 shows firefighters conducting marine rescue exercises in preparation for airplane crashes, capsized boats, and other offshore rescue situations (photo by author).

Fig. P–6 Firefighters Conduct Marine Rescue Exercises

Who would have predicted the development in the United States of 28 federal and state USAR task forces, each consisting of dozens of specially-trained firefighters, emergency room physicians, structural engineers, heavy equipment operators, logistics and communications specialists, hazmat technicians, and canine search teams? These units are strategically located around the nation, ready to be deployed within hours to earthquakes, floods, tornadoes, hurricanes, landslides, avalanches, and other disasters anywhere in the country. They are capable of locating and rescuing trapped victims around the clock for days or weeks.

Who would have predicted that selected U.S. fire department-based USAR task forces would become part of a system by which the State Department dispatches American rescuers to assist other nations conducting disaster SAR operations? These forces provide assistance in such far-flung places as Soviet Armenia, Greece, Turkey, Iceland, Greenland, Taiwan, the African continent, the Philippines, South America, the Caribbean, and other locations around the globe. Who would have predicted the evolution of an international system of SAR teams ready to converge from around the world to assist nations in dire need of assistance?

Who among the early fire service leaders would have predicted that the recruit academies of major municipal fire departments would devote weeks to urban search and technical rescue training? These academies have established a baseline requirement for new firefighters to enter the field with the basic skills required to pluck people from cliffs, extract them from collapsed trenches, tunnel them out of collapsed buildings, and pull them from flood waters.

These are all examples of the paradigm shift toward improved urban search and technical rescue capabilities. It's an exciting time for the fire and rescue services with respect to rescue and disaster response, and the opportunities for those interested in improving these capabilities are immense and ever-expanding.

A final intent of this series is to quantify both positive and negative lessons learned through years of experience by fire and rescue agencies with active USAR and technical SAR programs. The books specifically identify lessons learned *the hard way* by firefighters and rescue specialists assigned to active rescue companies, USAR units, swift-water rescue teams, helicopter rescue units, collapse teams, confined-space teams, mine and tunnel teams, mountain rescue teams, local/state/national USAR task forces, and other resources whose job is to manage urban search and technical rescue emergencies.

Defining urban search and technical rescue

Since the combined disciplines known as *USAR* broke onto the scene in the mid-1980s, a common question heard from citizens, politicians, and members of the news media alike is "What exactly *is* USAR?" One answer might be "A fire department-based capability that (depending on local hazards) includes such disparate disciplines as structure collapse SAR, confined-space rescue, river and flood rescue, high-angle rescue, mine and tunnel rescue, marine disaster, and helicopter rescue, etc." Another answer might be that USAR is the West Coast equivalent of "rescue" in the fire service. But in essence, it can be said that USAR includes any situation involving physical entrapment of humans (and, in some cases, even animals).

In Figure P–7, firefighters and lifeguards operate in tandem during a dive rescue operation after a recreational diver was pulled from the open sea into an offshore cooling water intake pipe and ended up in the bowels of a power plant.

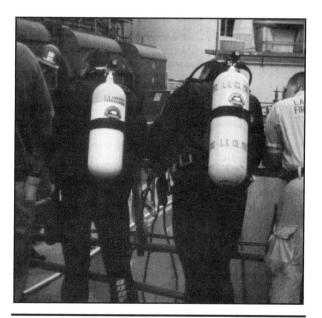

Fig. P–7 Diver was Pulled into the Freshwater Intake of a Power Plant

USAR, or if you prefer, rescue, also encompasses response to the consequences of terrorist acts such as bombings or sabotage, where victims may become trapped and require the assistance of technical SAR specialists. Finally, urban search and technical rescue encompasses disasters that typically result in trapped victims, including earthquakes, tornadoes, hurricanes, landslides, mud and debris flows, avalanches, and large-scale floods. It also includes effective management of rapid intervention emergencies in which firefighters or other public safety personnel find themselves trapped by structural collapse, disorientation in heavy smoke and/or darkness, or injury on the fireground. It also includes situations where rescuers require assistance (i.e. rapid intervention) during the course of non-fireground SAR operations.

Although USAR is most closely associated with metropolitan fire department operations, it's long been the author's observation that the word *urban* in USAR is somewhat of a misnomer. Many fire/rescue agencies that administer USAR programs are called upon on a daily basis to locate, treat, extricate, and transport victims in a variety of environments that don't resemble anything urban. Today, many fire departments around the world successfully manage technical SAR operations in wildland areas, in mountainous terrain, on the ocean, across the deserts, in lakes and ponds, along natural rivers and streams, in mines and caves, in resort areas, and even on glaciers and other locations that have no relation to urban terrain. Simply put, urban isn't an accurate description of the full range of environments within which modern fire/rescue agencies conduct SAR operations.

As more fire departments embrace USAR as an integral part of their mission—and as firefighters successfully complete ever-more-daring rescues with greater success—the term USAR will become more closely associated with the fire/rescue services in the eyes of the public and the media (Fig. P–8).

Fig. P–8 Fire Departments Acquire Advanced Extrication Equipment

For purposes of brevity, the conglomeration of disciplines discussed in this book series will be referred to as *rescue*. Clearly, not every fire department confronts this entire range of rescue hazards, nor is every fire/rescue agency responsible for managing every discipline within rescue. Local factors such as terrain, seismicity, geologic stability, age of buildings, previous and current construction practices, weather, demographics, mutual aid arrangements, and even the political climate (read: t.e.r.r.o.r.i.s.m.) generally dictate the range of emergencies and disasters for which fire departments should be prepared.

Recognizing tradition and adapting to new demands

As a life-saving service now demanded by a public that expects first-rate SAR service in practically every circumstance, rescue has become a valid, logical, and proven justification to expand the rescue capabilities of local fire departments. This is done by increasing staffing for engines, trucks, and USAR/rescue companies and USAR task force fire stations (where applicable), creating additional USAR/rescue units, and expanding fire/rescue agency budgets to provide better service to the public.

From the public's perspective, fire departments are generally the most cost-effective and capable entities prepared to provide advanced rescue service on a daily basis. As a core concept of fire and life safety protection, fire stations have always been strategically located to ensure rapid response times in the communities they serve. Fire departments have a long history of proven effectiveness in managing technical rescues and disasters. Add to this the fact that SAR is something firefighters have always done in varying degrees and in a variety of different environments, and one can see why rescue has become intractably entwined with the role of the fire service in providing the full range of emergency life safety services.

It's a mark of modern society that increasing degrees of specialization, backed up by specialized training, research and development, and resources, are required for fire fighters and rescuers to address the hazards with which they are confronted. Long gone are the days when the typical tailboard firefighter possessed the full range of skills to safely deal with all of fire and rescue hazards by virtue of traditional fire academy training and fireground experience mixed with good old-fashioned *horse sense.*

While it's generally acknowledged that common sense is among the most critical ingredients for successful emergency operations, sometimes common sense in itself doesn't provide all the tools needed to operate in *immediately dangerous to life and health* (IDLH) environments. Many firefighters have lost their lives doing what seemed logical at the time, without the benefit of formal training and experience about especially dangerous conditions like those found in many technical rescue emergencies and disasters. And many others (including this author) have had close brushes with death, surviving only through sheer luck or providence after being nearly overwhelmed during the course of physical rescues because they followed their instincts without the benefit of proper equipment

and without fully understanding the risks inherent to the hostile environments they were entering.

Examples abound of rescuers being lured into life-threatening predicaments during rescue operations by choosing seemingly common-sense courses of action that (in hindsight) were contraindicated by the conditions. One example is firefighters tying ropes around their bodies before attempting contact rescues in swift-flowing water, a tactic that regularly drowns would-be rescuers. Another example involves police, firefighters, and other would-be rescuers entering confined spaces without respiratory protection to attempt the rescue of downed victims—a potential death sentence in oxygen-deficient, oxygen-enriched, or otherwise toxic atmospheres. Other examples are rescuers entering unshored, partially collapsed trenches attempting to find buried construction workers and firefighters being suspended on the *upstream* side of bridges in ill-fated attempts to *pick off* victims being swept along fast-moving flood-control channels.

Another bad decision is using aerial ladders as cranes to lift victims in rescue litters without fully understanding the limitations of the ladder or the dynamic loads that can occur during rescue. Still another example is agencies adopting helicopter hoist cable systems for swift-water rescues without due regard for the tremendous dynamic loads that can impose on the cable *and the helicopter* by rescuers and victims attached to the cable in fast-moving water. Manufacturers even warn against this kind of use! Firefighters have been known to swim into the boil lines of low head dams to rescue victims without understanding the hydraulic forces that may quickly overpower them. And as a final example, police officers and firefighters alike have been pulled into the water while trying to snag victims from flood channels without personal protective equipment (PPE). The list of examples goes on and on, and each of us can probably think of one in which we or our colleagues were involved.

The element of risk can never be entirely eliminated from rescue, but it *can* be reduced through repetitive training, proper equipment, awareness of hazard, intelligent planning, competent incident command, teamwork, and (perhaps most important!) your own experience and learning from the experiences of others. Among these factors, experience and *realistic* training are among the most important.

This evokes the concept known as *recognition-primed decisions* (RPD), where ICs and other high-pressure decision-makers naturally and perhaps instinctually refer to their previous experiences in the course of making instantaneous decisions during dangerous and rapidly changing situations. Military studies have demonstrated that firefighters and other high-risk decision-makers subconsciously refer to mental *slide shows* gained from past experience that allow them to link their continuing size-up of the emergency with actions that have proven effective in the past. Thus, their decisions have been *primed* by previous experience. Decision-makers without vast emergency experience can benefit from realistic training and simulation, which replicates actual emergencies and primes them to quickly react to the real thing. RPD will be covered in greater depth later in this book.

Naturally, rescuers who are well-trained and thoroughly familiar with their equipment and the skills necessary to use it—for whom operating in unusual rescue environments is second nature—seem most capable of managing complex rescue emergencies in a safe and effective manner. In places where rescue incidents are less frequent, the next-best thing is to expose firefighters and other rescuers to realistic training and simulated conditions that replicate actual rescue environments.

Clearly, the immediate availability of fully staffed, highly trained, properly equipped units specializing in rescue, gives fire and rescue agencies a huge advantage when unusual rescues occur. And the best way to achieve this goal is through the use of (and proper support of) units that specialize in rescue, including dual-function companies that perform both fire and rescue functions (and particularly those that perform the role of *rapid intervention*.)

Figure P–9 shows how specialization in urban search and technical rescue is sometimes necessary for safe and effective management of unusual rescues like this dive rescue operation within the depths of an abandoned and submerged tunnel dug into a mountain in the 1800s.

Fig. P–9 Dive Rescue Operation in a Submerged Mine in the Mountains above Los Angeles

Most modern firefighters have the proven ability—and indeed the responsibility—to be prepared for a far wider range of emergencies than simply the consequences of burning buildings, wildland fires. In the modern world, the public has come to expect firefighters to be prepared for practically every non-law enforcement emergency that occurs in their communities, and in most cases their expectations are met or exceeded.

The Importance of Rescuer Safety

The final and perhaps most important goal of this book series is to reinforce the importance of firefighter and rescuer safety during the course of efforts to save the lives of missing or trapped people. As professional rescuers we're not in the business of blindly embarking on suicide missions, even though each of us may be willing to take extraordinary risks. We recognize that we may not return from the next mission if things go wrong. Although firefighters and rescue team members find satisfaction in the work they do, losing one's life doing what one loves tends to defeat the purpose *if the accident is preventable*. This point is illustrated poignantly in Mark V. Lonsdale's book *Alpine Operations,* in which he examines a recent spate of mountaineering tragedies resulting from the climbers' insistence on reaching the summit even when they're confronted with clear indications that they are over-matched by weather, altitude, and mountain, and that a safe return is doubtful.

Fig. P–10 Firefighters Practice High-rise SAR

> *For some the gamble pays off, but for many their destiny is to become a frozen lump on a wind-swept ridge, often not found until the following season if at all. The only eulogy being "He died doing what he loved." But, I ask you now, try to find me a climber lying in the snow exhausted and frostbitten, who says "Leave me alone to die, I am doing what I love."*

As Figure P–10 illustrates, rescuer safety must remain a top priority of modern fire/rescue departments. Realistic training is a critical aspect of rescuer safety. Here, firefighters practice helicopter-based high-rise firefighting and SAR operations from a helicopter shell mounted atop a three-story tower. The goal is to gain effectiveness and maximum safety through repetition of good technique, practiced under controlled yet realistic conditions.

The point is obvious enough; even though we may be willing to lay down our lives to save the lives of others, and although we signify that willingness each time we put on our respective badge, we must acknowledge that losing firefighters and rescuers to *preventable* accidents is an unacceptable outcome of modern rescue practice. And although untimely death comes more often than it should, the concept of planning for *acceptable losses* should not be part of our credo.

The loss of a highly trained and experienced rescuer not only affects his family and colleagues but may also have an adverse affect on other victims who might have benefited if that rescuer had lived to go on and do more good works in the future. The goal of professional rescuers is to get the job done but also live to pass on valuable information and conduct rescue operations in the future.

Far from being nebulous or euphemistic, all rescuers should consider themselves primarily *safety officers*, to make sure *they* don't cause an accident and to help prevent the line-of-duty deaths of their colleagues. Next, they should consider themselves *rapid intervention specialists,* ready at a moment's notice to spring into action to rescue a team member who's become lost or trapped. And, finally, rescuers should see themselves as *rescuers* for the victim. This idea is rooted in the most sacrosanct traditions of firefighting and rescue. It's a precept that's noted in a time-honored maxim that has become the working axiom of many teams: *First priority—save yourself; second priority—save your partner; third priority—save the victim.*

Over the years several of this author's colleagues have lost their lives in the course of rescue and firefighting operations. The author himself has been at the *zero-moment point* more than once while conducting *do-or-die* rescues. One such incident required a hastily organized rapid intervention operation to help save the victim—and myself—when conditions conspired to overwhelm us.

In addition to that, this author and the other members of his unit have been repeatedly tasked with rescuing other trapped firefighters on the fireground and during the course of rescue operations. Assisting a lost or trapped colleague or being trapped in the jaws of death yourself while attempting to save another person and looking up to see the faces of other firefighters scrambling to save you—knowing they may have your destiny in their hands—is a humbling experience. Such encounters tend to leave one with an intimate understanding of the importance of being able to rely on your colleagues when things go wrong and of the importance of having a proper rapid intervention system in place. These experiences also reinforce the personal responsibility that each of us has to remain constantly prepared to rescue a team member who finds himself in trouble. These events illustrate the criticality of research, planning, training, practice, and other measures required to ensure

adequate preparedness to assist our teammates, ourselves, and the victims who rely on us when the unexpected occurs.

Unfortunately, despite the current age of enlightenment about improving firefighter and rescuer safety through better training and equipment, we still encounter some fire/rescue officers, supervisors, and managers who *don't get it* when it comes to personnel safety and health. Some officers still allow (or direct) firefighters and rescuers to operate in untenable conditions even when there is no possibility of protecting lives, the environment, or property. Others fail to implement personnel accountability systems or rapid intervention measures when their personnel are exposed to IDLH conditions. Still others scoff at the special training and equipment requirements of those tasked with conducting urban search and technical rescue operations, preferring to rely on antiquated concepts of training and preparedness and ignoring proven concepts like RPD, which is covered later in this book.

The list goes on and on, and readers can probably think of examples to add to it. In this age of standards and information geared toward preparing personnel to manage high-risk operations with a reasonable degree of safety and the employment of rational risk vs. gain strategies, such lapses are inexcusable. The bottom line is that firefighters and other rescuers must remain alert to the dangers posed by emergency conditions, sometimes complicated by officers and supervisors who haven't yet bought into the concept that personnel safety is their primary responsibility.

Soviet high-altitude climber Anatoli Boukreev once described this philosophy. Boukreev was credited with single-handedly rescuing several climbers from a blizzard at 27,000 feet during a disastrous 1996 Mt. Everest expedition. Of high-risk operations, he said, "I come from a tradition that promotes mountaineering as a reasonable sports endeavor, not as a game of Russian roulette. The death of a team member is always a failure that supercedes any summit success."

As professionals whose job is to routinely per-form high-risk tasks, firefighters and rescue team members should adopt a similar credo: to come from a tradition that promotes the protection of life and the reduction of suffering as reasonable endeavors, not as games of chance; who believe in taking risks on a par with the threat to innocent lives; who are willing to risk their lives and the lives of their colleagues when other lives are at stake; and when they do, to reduce inherent risks through proper training, equipment, planning, and attention to basic safety principles.

Entrapment, injury, or death to firefighters or other rescuers during the course of emergency operations not only casts a pall over the emotions of those involved in the effort, but often redirects critically-needed resources from the rescue of citizens to rapid intervention operations. The preventable death of a rescuer during a successful rescue operation becomes a hollow victory.

For all these reasons, rescuer safety must remain a primary consideration during planning, training, and emergency operations. Urban search and technical rescue includes all forms of rapid intervention, a term given to the rescue of firefighters lost or trapped by collapse, disorientation, or injury on the fireground, as well as personnel trapped or lost in mishaps during non-fireground emergencies. Such emergencies can include cliff rescues, structure collapses, swift-water rescues, and trench rescues, etc. In many places, rapid intervention is the primary mission of fire depart-ment rescue companies, USAR units, and technical rescue teams.

If for no other reason, this text is intended to convey the message that firefighter and rescuer safety is *job number one* for those in charge of making decisions both before and during rescue emergencies.

Larry Collins

Hermosa Beach, California

November 2003

Acknowledgments

The information in this book was not developed in a vacuum, nor did it come to me in my sleep (with certain notable exceptions). In part, this book is the product of more than two decades of experience planning, instructing, conducting, and supervising a wide variety of technical SAR operations in the greater Los Angeles area, in other parts of the United States, and in several nations. The contents of this book are built on knowledge from many other firefighters and rescuers, instructors, authors, and researchers around the world.

Technical Rescue Operations is a humble attempt to pass on parts of an extensive (and ever-expanding) body of knowledge accumulated by many of the author's colleagues and friends in the course of careers dedicated to preparing for and responding to fires, technical rescues, and disasters.

The book is constructed on pillars of wisdom established by the likes of FDNY Deputy Chief Ray Downey (deceased), author of *The Rescue Company*. Deputy Chief Downey barely survived the collapse of the first World Trade Center tower but remained in the collapse zone to help other FDNY firefighters and officers rescue trapped people from the crushed Marriott Hotel until he was killed when the South Tower collapsed on all of them. This book also owes its origins to the works of authors like Slim Ray and Les Bechtel (*River Rescue*), Tom Vines (*On Rope*), Tim Setnicka (*Mountain Rescue*), Vincent Dunn (*Collapse of Burning Buildings*), Alan Brunacini (*Fire Command*), Francis Brannigan (*Building Construction for the Fire Service*), Chase Sargent (*Confined Space Rescue*), Jim Paige (*Street Dancer*), Bill Manning (Editor, *Fire Engineering* magazine), Mike Brown (*Engineering Practical Rope Systems*), Ronny Coleman (author of hundreds of educational editorials and columns), Mark V. Lonsdale (*Alpine Operations*), and the writers of other seminal fire/rescue service canon.

In *The Rescue Company,* Ray Downey provided the most definitive description to date of a modern fire department technical rescue unit. He helped quantify what it means to provide effective technical rescue services to a modern society. Others have followed in his footsteps, but rarely so effectively as Downey, who literally wrote the book on rescue. Ray was working on the second edition of *The Rescue Company* at the time of his death in the 9-11 terrorist attacks. That he was unable to finish the book and his other life's works is a great loss to the world's fire and rescue services and to the many victims of mishaps and disasters who ultimately would have benefited from the wisdom he communicated through his writing and instruction.

As a group, these pioneering firefighters, rescuers, instructors, and authors have been instrumental in identifying new challenges and solutions for the fire and rescue community. They codified many of the performance standards by which the fire/rescue services operate today. Their strong emphasis on improving the safety of firefighters and other rescuers has had the dual effect of improving the life expectancy of victims, for we know that it's an inherent truth that safer emergency operations are directly related to factors like greater efficiency, better equipment and training, and superior command. Each of these factors, in turn, directly benefits victims trapped in building collapses, floods, and other predicaments requiring physical rescue.

This book strives to build upon and extend the reach of the previously-mentioned authors by (1) providing a new perspective on traditional rescue challenges, (2) offering a broader view of what it means to be a USAR/rescue company (or USAR task force), and (3) describing systems-based approaches to managing daily technical rescues and SAR operations resulting from man-made and natural disasters.

In addition to the individuals mentioned above, the author credits the following pioneering programs for helping lay the groundwork for a new era in rescue: the USAID and the OFDA which coordinate American-bred international USAR response capabilities in a cooperative effort with USAR task forces based in the Miami-Dade (FL) Fire and Rescue Department, the Fairfax County (VA) Fire and Rescue Department, and the LACoFD; FEMA's trailblazing system of 28 USAR task forces and the related training courses that continue to be developed and taught today; many state offices of emergency services (including California OES) that have advanced the cause of USAR, swift-water rescue, and ICS; the California State Fire Marshall's seminal Rescue Systems I and II and Emergency Trench Rescue courses, which have been adopted nationally; swift-water/river and flood rescue courses developed by the Ohio Division of Natural Resources; the Albuquerque (NM) Fire Department; the Los Angeles County Multi-Agency Swift-water Committee; and various public and private organizations.

I also acknowledge the general body of knowledge behind standards like NFPA 1670 (*Standard for Technical Rescue*), NFPA 1500 (*Standard for Firefighter Health and Safety*), OSHA's *Two-In, Two-Out* rule, and various other worker safety standards and regulations and the INSARAG, which operates under the umbrella of the United Nations, whose mission is to develop effective international relationships to save lives and render humanitarian services following natural or man-made disasters, to improve emergency preparedness, cooperation, and information exchange between international SAR teams, and to develop a common understanding of the functions and operations of SAR teams. The result of this effort is the development of the International SAR Response System.

Notable among the true experiences that helped set the stage for this book and those of the other agencies and organizations include the FDNY, the Albuquerque Fire Department, the LACoFD, the Montgomery County (MD) Fire Department, the Virginia Beach (VA) Fire Department, Rescue Three International, the London Fire Brigade, the Ventura County (CA) Fire Department, the Orange County (CA) Fire Authority, the San Diego (CA) Fire Department, the Istanbul Fire Brigade (Turkey), Spec Rescue International, the New South Wales (Australia) Fire Brigade, the Melbourne (Australia) Fire Brigade, the Dallas (TX) Fire Department, the City of Los Angeles Fire Department (LAFD), the Specialized Tactical Training Unit (STTU), the International Emergency Technical Rescue Institute (IETRI), the Dayton (OH) Fire Department, the Indian Wells (CA) SAR Team, the SMART Team of Pennsylvania, the Lifeguard Division of the County of Los Angeles Fire Department, the Oklahoma City (OK) Fire Department, and the Los Angeles County (CA) Sheriff Department's volunteer Mountain SAR Teams and Emergency Services Detail.

Over the years this author has had the privilege of being associated with a venerable cast of individuals who (as a group) embody a sort of *who's who* of rescue. The names are far too numerous to list in total, but here is a brief sampling of those whose influence can be seen in this book: Mike Tamillow, Mark Ghilarducci, John O'Connell, Mickey Conboy, Geoff Williams, Dewey Perks, Carlos Castillo, Ruben Almaguer, Jim Hone, Mike McGroarty, Jim Seigerstrom, Jim Mendonza, Warwick Kidd, Trevor Owens, Buddy Martinette, Joe Barbera, Dave Hammond, Chuck Mills, John Huff, Joe Brocato, Jeff Frazier, Rob Patterson, Dave Downey, John Oseteraas, Tim Gallagher, Bob Samuelson, Chuck Mills, and many others.

Naturally, many other individuals and agencies are deserving of thanks and recognition, including my colleagues and mentors of the past two decades who contributed so much of their knowledge and experience. Chief among them are Mike Minore, Keno DeVarney, Wayne Ibers, Don Roy, Reggie Lee, Bill Masten, Mike Layhee, Frank McCarthy, Ysidro Miranda, Rick Meline, Leo Ibarra, Brian LeFave, Dan Gordon,

Mike Inman, Dave Norman, all rescue pioneers and original members of USAR-1, the LACoFD's original USAR company. The list includes other rescue pioneers like Mike Erb, Don Hull, Don Tayenaka, Rick Cearley, Jeff Langley, Carl Cotton, Bob Dunbar, Tom Short, Dan Fourniete, Lee Benson, Joe Moline, John Haugh, Pat Rohaley, Rory Rehbeck, John Boyle, Frank Pratt, Rich Atwood, and Vern Atwater, all of the LACoFD.

My sincere gratitude rests with the following people whose professional missions led us to cross paths and whose contributions to rescue cannot be denied: Peter Henderson (USAID/OFDA); Congressman Howard Berman (California); Doug Campbell (International Affairs Aid to Congressman Berman); Costas Synolakis (Professor of Engineering at USC and world-renowned expert on tsunamis); Professor Tom Henyey (Director of the Southern California Earthquake Center [SCEC]); Mark Benthian and Jill Andrews (SCEC); Kerry Seih, Thomas Heaton, and Lucy Jones (all seismologists at Cal Tech/SCEC); Barry Perrou (L.A. County Sheriff Department Crisis Negotiation Team); Dennis Smith (author, *Report From Engine Company 82* and *Report From Ground Zero*); Marla Petal (Project Manager, *Istanbul Community Impact Project*); and the Los Angeles County Board of Supervisors, whose members consistently supported the development of expert USAR capabilities in L.A. County, across the United States, and internationally for two decades.

Also deserving of mention for their leadership roles in the rapid evolution of urban search and technical rescue and its application in daily fire service operations are Fire Chief P. Michael Freeman, Chief Deputy Larry Miller, Chief Deputy Gary Lockhart, Deputy Chief Jim Ryland, Deputy Chief Mike Dyer, Deputy Chief Daryl Osby, Battalion Chief Jim Powers, Battalion Chief Terry DeJournett, Assistant Chief Jim Sheppard (retired), Battalion Chief Don Oldham (retired), Assistant Chief Marc McConnel (retired), Assistant Chief Mike Bryant, Assistant Chief Mike Morgan, Assistant Chief Ron Jones (deceased), Deputy Chief William Zeason (retired), Fire Chief John Englund (retired), Fire Chief Clyde Bragdon (retired), Deputy Chief Ronald King (retired), and the current administration of the LACoFD. Their visionary support of urban search and technical rescue since the early 1980s led to a vast and pioneering effort to improve the manner by which trapped people are rescued in L.A. County, in the state of California, across the United States, and abroad. They provided my colleagues and me with the opportunities by which we have gained invaluable knowledge and experience in this dynamic discipline, the ultimate result of which has been the saving of lives that otherwise would have been lost.

This book would be less than complete without mention of several colleagues who were provided inspiration and made valuable contributions to the fire and rescue services before losing their lives in the line of duty.

Fire Fighter/Specialist (deceased) Jim Howe personified the hardworking, intelligent, humble, dyed-in-the-wool firefighter upon whom the most positive stereotypes of the fire/rescue profession are based. He dove into emergency operations with such ferocity that he was for years known by the nickname *Tasmanian*. And although he was known for his wicked sense of humor, Jim Howe was a deeply religious family man. I personally owe Jim Howe a debt of gratitude for once reaching out with his fire axe to yank me from a hole created by my body when I fell partway through the roof of a burning apartment complex. Seconds later, standing on the ridgeline while I checked myself for burns, still wide-eyed from my brush with disaster, Howe's only comment before we went back to work was "Close call, kid."

Five years later, the tile-laden facade of a burning multi-story office complex peeled off the building, fatally burying Howe and trapping several other firefighters beneath tons of debris on a second floor landing. Images of Jim Howe's helmet visible within the collapse pile—as well as the other firefighter fatalities—helped spur the implementation of formal rapid intervention protocols in the LACoFD and other fire/rescue agencies in the West (following the example established by many east coast departments) long before they were mandated by the Occupational Safety and Health Administration (OSHA).

As a rookie in the Ventura County (CA) Fire Department, Fire Fighter (deceased) Gordon Daybell was a young, smart, well-educated, aggressive firefighter and colleague of mine who diligently prepared himself for a long and productive life in public service. His good deeds and his drive to help people served as a reality check for his colleagues.

One night a citizen banged on the station door to report that a car had skidded off the road into nearby Lake Piru. The crew of engine 13 arrived on the scene of the accident to find a pair of headlights dimly visible from beneath the frigid and murky water. Without hesitation, Gordon Daybell stripped off his turnouts and made repeated dives without the benefit of scuba or snorkel and mask. It was a dangerous situation because the car had *turtled* and buried itself upside down in the muddy bottom. Daybell was finally able to break out a window, free the driver, and bring him to the surface. After swimming the victim to shore, Daybell and the other firefighters initiated CPR on the pulseless man. The driver lived, walking out of the hospital just days later, suffering minimal deficit.

Within weeks of the dramatic rescue, Daybell began suffering unexplained numbness in his hands, accompanied by severe headaches. He was soon diagnosed as having an inoperable brain tumor unrelated to the incident in Lake Piru. Rapidly debilitated by the effects of the fast-growing tumor, Gordon died within two years. Considering Gordon Daybell's heroics at the lake, a friend later remarked that God had traded a life for a life. It's sometimes difficult to understand such an equation, considering the number of lives Daybell might have saved had his life not ended at such a young age. In the end, much about life and death remains a mystery to us.

Fire Fighter/Paramedic (deceased) Jeff Langley was a man to whom many people owe their lives and yet don't even know it. Langley made invaluable contributions to public safety and firefighter survival before he fell to his death during a helicopter-based cliff rescue operation in Malibu in 1993.

As a firefighter/paramedic assigned to the air operations section, Jeff Langley was part of a group of LACoFD personnel who pioneered a daring series of methods and new rescue equipment now used by helicopter-based, swift-water rescue teams worldwide to pluck people from fast-moving flood-control channels and raging rivers. In southern California alone, dozens of people owe their lives to these innovations. The list will continue to grow worldwide as the years creep on. Langley was a constant advocate of improved safety and effectiveness in helo-rescue operations. The National Association for Search and Rescue's annual Langley/Higgins Award was named in honor of Langley and Earl Higgins, a citizen who lost his life attempting to save a child from the raging Los Angeles River. The tragic irony of the manner in which Langley lost his life has not been lost on those who knew and worked with him.

Introduction

Rescue is among the most challenging and dangerous tasks faced by firefighters and others charged with the protection of life. Rescue is a broad-based mission that changes in concert with the risks of our social, industrial, natural, and political environments. For many, the increasing challenge and danger of modern rescue was crystallized on 9-11-01 when terrorists brought down four airliners in three states and collapsed part of the Pentagon, causing the two largest and most deadly building collapses in history. The 9-11 tragedy killed more than 400 firefighters and police officers and thousands of citizens and changed the course of history. As you read this book, terrorism remains an ever-present threat that is likely to present far more dangers to civilized societies (and those charged with protecting them) in the coming years. In addition to other rescue-related hazards, which are considerable and growing every year, the public expects firefighters and rescuers to respond with efficiency and clarity of purpose to terrorist attacks.

Rescue is a profession, a science, an art, and a calling. The scope of rescue is rapidly expanding with diverse disciplines such as helicopter rescue, mountain rescue, underground rescue, confined-space rescue, water rescue, USAR, terrorism-consequence management, and others under its umbrella. In many industrialized nations, rescue has been recognized as a standard role of firefighters and other public safety members. The public has come to associate timely and professional rescue with the fire service and various rescue organizations. Today, first responders from local fire and rescue agencies are expected to manage a wide range of *daily* rescues, and local ICs are expected to request and use specialized rescue resources in a timely and effective manner when their assistance becomes necessary.

Locating and extracting missing or trapped victims is among the most critical jobs in public safety. Successful rescue operations require dedication to protecting lives, a willingness to take calculated risks, and an understanding of (or ability to figure out) how things work… and, consequently, how things fail. It requires the ability to devise and implement solutions for complex problems under hazardous and rapidly changing conditions, all without the luxury of adequate time to research and digest all the information one might normally want before committing personnel to high-risk conditions. Rescue is performed during the most crucial moments of peoples' lives by firefighters and other rescuers who are guided by instinct, experience, protocol, and the ability to improvise and take necessary risk.

The essence of rescue was demonstrated by the FDNY and Washington D.C. fire departments in the harrowing minutes and hours following the 9-11 terrorist attacks. The firefighters, police officers, EMS personnel, and other rescuers who rushed into the World Trade Center towers and the Pentagon exemplified the very spirit of rescue. Many of them crossed the proverbial Rubicon, never to return. But they did so willingly, placing the lives of others ahead of their own, using whatever resources were available to them under harrowing conditions and never shrinking from their duty to protect the public. Never has there been such a clear display of devotion to duty. And never before has there been such a clear example of why rescue is at the very core of what firefighters and other public safety personnel do.

Just as the Chinese word for *crisis* also symbolizes *opportunity*, *rescue* represents both good and bad fortune hanging in the balance. Rescue emergencies separate those who are prepared to make decisive moves at the right time with the right equipment and training, supported by effective systems, from those who are not. For personnel faced with the task of rescuing victims lost or trapped in desperate circumstances, rescue becomes a sort of *crucible*, a test not only of their training, experience, and planning, but also of their resourcefulness, daring, and ability to adapt to rapidly deteriorating conditions when lives are on the line. It is among the purest demonstrations of risk vs. gain and of the willingness to risk life to save the lives of others. Few activities match the level of human drama, and fewer still carry such lethal consequences for the unprepared or the unlucky.

Although rescue is accomplished through widely varying strategies, tactics, and regional styles, there's little ambiguity in the bottom line. Victims either live, or they do not; they are either located and rescued, or they are not. Few outcomes are as clear as in rescue emergencies, where the level of preparedness of the rescuers is painfully evident to all.

It's not by serendipity that fire/rescue officers are studied by military officers and behavioral scientists interested in observing how time-critical decisions are made under true life-and-death conditions. Municipal fire department officers were the subjects of seminal studies in RPD conducted by the U.S. Army in the 1990s. Fire and rescue emergencies have proven ideal for the study of humans who must solve complex problems in potentially lethal environments with severe time constraints. In terms of time-compressed decision-making, and risk vs. gain decision-making, rescue and firefighting are considered a close approximation to the battlefield.

Consider for example the sequence of events necessary for the successful conclusion to a typical rescue emergency. Firefighters and other rescuers must first arrive on the scene with sufficient resources to begin initial action. They must accurately recognize the victim's predicament and its most critical components within seconds of arrival on the scene. They must then sort through the chaos of the emergency scene to understand the essence of the problem and devise solutions (with alternate backup plans) based on the conditions of the scene and resources that will arrive at lengthening intervals with varying levels of equipment and experience. Next, they implement the operational plan with surgical precision, knowing that the smallest mistake may result in instant death to victims or rescuers. Then they must be ready to evaluate the effects of their strategy *while their plans are being implemented*, redirecting tactics without delay to rescue the victim without creating casualties among would-be rescuers.

The typical rescue scene is fraught with dangers and conditions that may affect the final outcome before they're even recognized. The process of rescue is complicated by the pressure of knowing that lives may be won or lost based on the decisions, physical stamina, and the willingness of rescuers to risk extreme physical danger. Rescue is where many fire/rescue professionals really earn their money: when they find themselves in the midst of chaotic, rapidly changing situations with lives on the line, and they must quickly decide on a course of action to mitigate the problem and remove victims from harm without being injured or killed themselves. If the primary role of the fire/rescue services is protection of life, rescue is where it happens. For purposes of this book, the category *fire/rescue professionals* includes members of volunteer fire departments and SAR teams.

The ancient Roman fire brigade (the *Militia Vigilium)* used certain tactics, strategies, and equipment that were admittedly crude by today's standards. Firefighters in places such as Constantinople, London, Prague, and other Old World cities were certainly called upon to rescue people (and even animals) trapped in burning buildings and other predicaments. Nevertheless, it took centuries for rescue to be identified as a discipline requiring dedicated units whose members were specifically trained to extract people from difficult entrapments.

In the early 20th century, a number of U.S. fire departments on the east coast established dedicated rescue companies to help ensure the timely rescue of firefighters who became lost or trapped during the course of fireground operations. According to Ray Downey in *The Rescue Company*, in 1915, the FDNY implemented a new concept that eventually became known as a rescue company. New York's Rescue Company 1 was the first unit in the Americas whose primary role was to conduct specialized rescue operations and to provide heavy rescue capabilities at the scene of difficult fires. Staffed by a captain, a lieutenant, and up to eight firefighters, Rescue Company 1 set the standard for a new approach to rescue and fireground operations. In 1917, the Boston Fire Department established its own rescue company, equipped and staffed to conduct the most difficult rescues and to give fireground commanders unprecedented options for forcible entry, fire attack, and the rescue of downed firefighters.

In 1929, the Chicago Fire Department followed suit with the creation of three heavy rescue companies. These pioneering agencies quickly demonstrated the efficacy of specialized rescue units whose primary job was to handle high-risk rescue operations, including what (nearly a century later) has become known as *rapid intervention* (the rescue of fellow firefighters who become lost, trapped, or otherwise incapacitated on the fireground and other emergency scenes).

Today, rescue companies and their progeny, USAR companies, USAR task forces, squad companies, swift-water rescue teams, USAR truck companies, and USAR teams are key rescue resources for many progressive fire/rescue agencies across the United States and many other nations. These units provide a method by which highly trained personnel can be concentrated in one place and given years of experience handling rescue emergencies. The lessons learned can be identified, documented, and passed on in the form of training to other rescue professionals, thereby creating a deep and vast pool of knowledge and experience from which to draw in times of complex rescue emergencies and disasters (Fig. I–1).

Fig. I–1 Modern Rescue/USAR Companies Perform Full Range of Rescues (Courtesy Gordon Massingham)

Without dedicated rescue companies, USAR units and other companies whose members are charged with experiencing, reviewing, quantifying, teaching, and implementing the lessons from repeated rescue operations, some fire/rescue agencies are forever doomed to repeating the same mistakes. They lose valuable experiences to first responders whose job doesn't include documenting and implementing the lessons and maintaining a force that's less then optimally prepared to manage the rescue emergencies and disasters that affect modern society.

While this book touches on the operation of specialized units like rescue companies, USAR companies, squad companies, and various types of SAR teams (including regional, state, national, and international USAR task forces), it's important to recognize that the vast majority of firefighters and rescuers in the world work for volunteer and professional agencies lacking the fiscal resources to establish and maintain dedicated rescue units. That is why the principles, preparations, strategies, and tactics in this book also apply to the typical fire department truck company, as well as engine companies, brush patrols, fire/rescue helicopters, chief officers, disaster planners, rescue instructors, and others responsible for managing urban search and technical rescue emergencies and disasters (or those charged with teaching others how to do it).

Therefore, this book is also intended as a reference for city managers; emergency managers; local, state, and federal government officials; and others whose decisions directly or indirectly affect the outcome of rescue operations and disasters. It can even be used as a means of educating members of the news media, who in turn keep the public informed about the manner and effectiveness by which fire/rescue agencies do their work.

Unique Factors Associated with High-Visibility Rescues

The preparedness of local fire/rescue agencies confronted with difficult rescue problems is closely scrutinized by the news media, politicians, and the public. In rescue, success or failure is usually immediate and final. The victim either survives or does not. In the case of earthquakes and other disasters resulting in rescue operations lasting days or even weeks, the outcome may not be immediate, but it's every bit as final. It's becoming increasingly common for entire rescue operations to be broadcast live to the public in *real time* on their television sets. The modern reality is that local preparedness and response are also closely watched by terrorists intent on conducting attacks. Modern terrorists will often evaluate local fire/rescue response to identify shortcomings and to capitalize on them.

In today's workplace environment, many high-visibility emergency incidents are also being watched by representatives of workplace safety agencies. Workplace safety inspectors are increasingly less likely to overlook infractions, especially when firefighters and other would-be rescuers are injured or killed. But this shouldn't necessarily be construed as a negative factor. The wise IC or rescue officer will often welcome the presence of a knowledgeable and helpful worker safety representative at the scene of a difficult rescue. In fact, it's become this author's practice to request a local worker safety representative to respond directly to the scene of certain lengthy or complex extrication operations such as trench collapses, confined-space accidents, and structural collapses. Not only does this aid in documenting the cause of the accident, but the representative can be used as a technical specialist and consultant to review the efficacy of ongoing rescue operations and advise if problems have gone unnoticed. In southern California, as in other states, the state's OSHA agency is the local workplace safety authority. California OSHA (Cal/OSHA) dispatches a local investigator to the scene of serious workplace injuries. A local Cal/OSHA investigator will also respond to the scene at the request of the fire department, not only to identify the cause of a workplace accident but also to give advice to firefighters and rescuers during the course of rescue operations.

The worker safety representative can also be assigned to advise the incident safety officer on matters related to firefighter/rescuer safety. This has the effect of providing the worker safety representative with a view of the incident and its many complications and hazards from the perspective of the fire/rescue agency, which may avoid a situation where he or she later becomes a potential adversary. It's this author's experience that the fire department's emergency operations are almost always portrayed in a very positive light in the official industrial accident reports prepared by the worker safety representative. In terms of liability for ICs and company officers, this is an important consideration for the future.

Another undeniable element in this process is hard to ignore. A budding army of attorneys with savvy about the intricacies of emergency operations have taken to observing fireground and rescue operations with a critical eye. Gone are the days when the fire/rescue services had a sort of invisible immunity from the scrutiny of the legal system. The veil of immunity has been forever lifted by several landmark cases in which the tactics and strategies of fire/rescue agencies have been called into question. This trend can be expected to accelerate as lawyers discover that mistakes on the emergency scene—or acts and omissions during the planning and preparedness process—may represent a sort of cash cow for those willing to exploit them. This is yet another reason for fire/rescue agencies to maintain a positive working relationship with investigators from worker safety agencies. Many fire/rescue agencies have a polity to immediately notify the district attorney's office when serious workplace injuries or fatalities are encountered (Fig. I–2).

Fig. I–2 Modern Fire Departments Must Maintain Constant Readiness (Courtesy Gordon Massingham)

Equally important is the moral and legal obligation of fire/rescue agencies and their employees to maintain constant readiness to provide timely, reasonable, and professional response when citizens become lost or trapped under conditions that require rescue tactics and strategy. Given the strategic location of fire stations throughout the industrialized world, as well as the vast amount of information, equipment, training, and mutual aid available—not to mention the precedence established by other fire/rescue agencies with similar hazards—it is inexcusable for local fire departments to show up at the scene of emergencies and disasters unprepared. It would be inexcusable to be without a

plan and without backup plans that include timely requests for mutual aid or state and federal assistance when appropriate. And it's inexcusable in the modern age for any fire/rescue agency to pass off rescue as *something we don't do*.

Despite the media draw of complicated and lengthy search rescue operations that sometimes place rescuers in the public eye, the reality is this: Rescue is often a dirty, labor-intensive, dangerous, and decidedly unglamorous process. Training and preparing for effective rescue is even dirtier and more labor-intensive, and far less glamorous.

When emergency operations are conducted by personnel who lack basic technical SAR skills or proper equipment or, for whatever reason, are generally unprepared for the complex demands of rescue, things can quickly go wrong, sometimes with devastating results. Conversely, rescue can be one of the most rewarding activities when fire/rescue personnel are properly trained, equipped, and adequately prepared to deal with the consequences of local and regional hazards. Even when unforeseen misfortune complicates SAR efforts, properly prepared firefighters and rescuers are ready to implement alternative solutions without missing a beat. This is where well trained and highly disciplined rescue companies, USAR task forces, and other rescue resources make the difference in the outcome of difficult SAR operations. It's where many of us truly earn our money.

The New Face of Rescue

The variety of rescue hazards has never been so diverse. Today there are urban fire department rescue units in some regions whose members might find themselves being dispatched to a high-rise fire or an industrial fire *in the city* in the morning, then to a cliff rescue or a plane crash in a 10,000-ft mountain range on the very next response (and this is particularly true of USAR and rescue units in those departments). Conversely, there are rural fire departments and mountain rescue teams that can be requested to assist with tunnel rescues or trench collapses in towns and suburban areas. The lines separating urban and rural fire/rescue departments are increasingly blurred by the advance of better SAR-related technologies, new rescue capabilities, and expanding rescue-related responsibilities (Fig. I–3).

Fig. I–3 Modern Fire Departments Conduct Mountain Rescue and High-angle Rescue Operations

Around the world, the fire service has stepped up to assume responsibility for ensuring safe, timely, and effective response to daily rescue-related emergencies. Today, fire and rescue departments handle the most complicated and dangerous technical SAR emergencies almost as a matter of course. Not so long ago, many more victims of such mishaps and disasters died without being rescued, in part because few firefighters and rescuers were equipped or trained to manage the consequences in a timely manner. Today it's become rather routine for victims of seemingly non-survivable mishaps to be rescued alive and returned to health because firefighters and other rescuers have the knowledge, experience, skills, and equipment to locate, access, and treat them while they're still trapped. Rescues that once qualified as *miraculous* now are often viewed as the expected outcome of intelligent planning, effective training, and good overall preparedness by local fire departments and other rescue agencies.

Almost without exception, firefighters are the first line of defense for civilized nations, and not only when fires occur. They are often first on the scene of medical emergencies, hazmat releases, natural disasters, and man-made disasters like terrorist attacks. Firefighters are part of the first wave of rescuers when people are trapped on cliffs, in collapsed trenches and excavations, in machinery, within collapsed structures, in swift-moving water, beneath avalanches and landslides, within mud and debris flows, in transportation accidents, and innumerable other predicaments. Today the general public recognizes that firefighters are often responsible for managing the most critical moments of emergency incidents, when time-critical decisions and actions have the most lasting and profound effect on the lives of affected victims, setting the stage for all the consequence management operations that follow.

In places where dedicated fire department rescue companies or urban search and technical rescue programs have been in place for years, technical rescue has long been considered a priority. In these places, local rescue capabilities have been maintained at a high level of performance almost as a matter of course. But in many other regions, rescue has rapidly evolved only during the past two decades, radically changing the face of the fire service in those places. New USAR and rescue programs have appeared practically out of nowhere in fire departments whose members for decades eschewed anything resembling formal rescue response systems. In addition, many existing rescue systems have been reinvigorated by the newfound emphasis on technical SAR and disaster rescue. In short, the trend toward expert management of rescue has been nothing short of a revolution.

Jurisdictional and Responsibility Issues

This book series purposely avoids making distinctions between agencies that respond to USAR and municipal rescue emergencies, those that handle mountain rescue situations, those that cover desert or coastal terrain, and those that service other environments. Such distinctions serve little useful purpose here because a growing number of modern fire/rescue departments are responsible for managing emergencies in all these environments, and it's becoming ever-more common for fire/rescue units to respond to them practically on a daily basis. In the best-case scenario, all agencies responding to rescue emergencies are using the ICS and have developed cooperative agreements or understandings that allow the responders to work closely with one another with minimal conflict. Optimum cooperation and coordination is more important than ever in the face of the growing threats to emergency responders.

That said, the issue of jurisdictional responsibility is worthy of critical review because it has become a sticky issue in some regions. In some places, the advent of fire department-based rescue (or urban search and technical rescue) capabilities is causing conflict between those who traditionally managed rescue, and firefighters who are sometimes equally trained, equipped, and experienced, and who may often arrive sooner. Conflict has already occurred with respect to technical SAR operations in some places where non-fire teams assumed that they held jurisdiction, but have been surprised to arrive at the scene to find firefighters already engaged in technical SAR operations. Fire departments frequently are responsible for certain SAR operations in mountain areas. In some places, conflicts have arisen out of these situations, with some traditional SAR team members arguing that fire departments shouldn't be conducting technical rescue operations in the

mountains and wilderness areas, even when well-trained and equipped firefighters are on duty 24 hours a day in fire stations located closer. It's a dilemma that's also been noted in relation to dive rescue, mine and tunnel rescue, trench rescue, and other rescue disciplines. Clearly there are a number of turf wars brewing over who has jurisdiction over rescue emergencies, and the effects are far reaching. But it's safe to say that the dynamics of rescue are changing in many regions.

In Figure I–4, a USAR company advances up a mountain road to reach a neighborhood that's been struck by mud and debris flows.

Fig. I–4 USAR Company in a Mud and Debris Flow

By virtue of strategic positioning of fire stations for timely response to daily fire, EMS, and other emergencies, local fire departments are often first on the scene of rescue emergencies, whether they occur in the city, in the desert, on the ocean, in the mountains, or elsewhere. The public has come to expect the local fire/rescue department to properly manage these emergencies. And if the local fire department finds itself overwhelmed by the scale or complexity of a particular rescue or disaster, the public expects the chief officers to request appropriate assistance without delay to mitigate the hazard with minimum loss of lives.

In short, rescue (both urban and wilderness) has become a primary function of many progressive fire/rescue agencies. Today, few things are more fundamental for modern firefighters than maintaining constant readiness to properly manage the search for, and the rescue of, victims. This includes the rescue of other firefighters in *rapid intervention* situations that can develop during the course of fireground operations and rescue-related emergencies and disasters. The trend toward expanding the fire service role in USAR and technical SAR, as well as other specialty fields, shows no sign of slowing down anytime soon.

Given these facts, it's also important to stress that by no stretch of the imagination is rescue solely the function of fire departments. There are many allied SAR organizations and teams across North America that continue to provide the highest levels of service to people lost, trapped, or otherwise stranded in precarious situations requiring physical rescue. Even the military plays a role in these efforts at times. And the public (for good reason) is becoming less tolerant of SAR operations that are bungled because of inter-organizational disputes. It's no longer acceptable (and never should have been palatable) for rescuers to argue over jurisdictional boundaries and responsibilities while innocent people await their assistance.

These inherent factors not only place the fire service in a leadership position, but they also place great responsibility on local fire departments to do the job right when called upon to locate and extricate trapped victims. Thus, there is an ever-expanding need for expert SAR capabilities within the fire service, accompanied by an increasing emphasis on fire department rescue companies, USAR companies, USAR task forces, and other technical SAR resources.

Although disciplines like USAR, technical SAR, and swift-water rescue are sometimes seen as new landmarks on the fire service landscape, rescue has always been—and always will be—closely associated with the fire service. Despite the objections of some traditionalists who bemoan the advent of fire department-based EMS, hazmat teams, and now rescue (or USAR), the fire service has generally embraced these expanded roles as part of their baseline responsibilities to public safety and service. And rightly so. The public (not to mention the news media and politicians) has come to demand prompt, professional technical SAR service from its fire departments.

There are, of course, a variety of professional and volunteer rescue service agencies that also provide timely and expert technical SAR service. In some regions, these agencies are clearly the most highly qualified to take the lead role in such operations, with the fire department providing support. The main component is good cooperation among all responders, and an understanding and recognition of the strengths and limitations of each agency. For those reasons, *Rescue* wasn't written strictly for fire departments, but also for members of allied rescue groups and teams.

New Paradigms

Recognizing that the public's perceptions and expectations (not to mention new laws and standards) have become a sort of mandate, many fire/rescue department administrators have concluded that firefighters should (or must) be proficient. They must be proficient in the skills required to conduct SAR in practically any local environment that creates the potential for dangerous rescues. This represents nothing short of a paradigm shift. The use of new tools requires extensive training to maximize effectiveness, to understand limitations, and to determine the range of applications (Fig. I–5).

Fig. I–5 New Tools Require Extensive Training

To cite specific examples of the effect of this paradigm shift, one may point to the following factors:

- the emergence of urban SAR as a recognized discipline within the fire/rescue service

- the rise of swift-water rescue as a necessary firefighter skill and the rapid evolution of high-angle rescue techniques and equipment

- development of the seminal Rescue Systems and Emergency Trench Rescue courses, which have been adopted nationally in the United States and are being taught in other nations

- the enacting of new laws governing confined-space rescue, not to mention the corresponding effect of OSHA's *two-in, two-out* rule on interior structure firefighting and other fire department operations in atmospheres identified as IDLH

- new NFPA standards on firefighter safety and health and rescue

- the expanding fire service role in improving the effectiveness and timeliness of mountain SAR operations

- the role of the fire service in anti-terrorism-related consequence management

- the public's growing awareness of local fire departments as a primary provider of technical SAR service

Disaster Rescue and the Widening Role of Fire Departments and State/Federal Rescue Resources

The need for effective coordination between local, state, and national rescue resources is especially important when disaster strikes. During the past two decades, U.S. fire/rescue agencies—often supported by state offices of emergency services and federal agencies like FEMA—have successfully managed rescue-related consequences of disasters. These disasters included earthquakes, tornadoes, floods, hurricanes, landslides, mudslides, avalanches, terrorist bombings, and a host of other disasters that trapped people in collapsed buildings, beneath debris, at heights, and other predicaments. Firefighters are often first on the scene of disasters, and they are among the last to leave.

The 9-11 terrorist attacks proved conclusively that fire departments have been—and will always be—the first and most critical first line of defense against most types of disasters. Increasingly, the rescue resources established by states and nations are playing a pivotal support role in the expert management of disaster-related SAR operations, especially those that continue around-the-clock for days and weeks.

The rise of USAR as a priority of the U.S. federal government in the wake of disasters like the 1985 Mexico City earthquake, the Armenia quake, Hurricane Hugo, and the Loma Prieta earthquake in 1989 marked yet another sea-change in the way technical SAR operations are managed. The Miami-Dade (Florida) Fire and Rescue Agency (already teaching disaster preparedness/rescue in the Caribbean and Latin America) and firefighters from several California fire departments and the California Office of Emergency Services (OES) were all dispatched to the Mexico City earthquake. Afterwards, the need for formalized teams that could be deployed by the U.S. government to assist nations faced with difficult SAR problems in collapsed structures and other disaster sites became apparent.

Taking a pro-active approach to the growing problem of urban disasters across the globe, the USAID/OFDA developed a partnership with the Miami-Dade Fire and Rescue Department to establish just such a team. Through their pioneering efforts responding to other disasters on foreign soil, the USAID/OFDA/Miami-Dade team became the forerunner of what would eventually become a national system of USAR task forces across the United States.

Several years later, as a result of innovative preparedness work with the Iceland fire service and the need for international rescue resources at events like the earthquake in Armenia, the USAID/OFDA system was expanded to include an internationally deployable USAR task force based in the Fairfax County (Virginia) Fire and Rescue Department. In the ensuing years, the USAID/OFDA teams from Miami-Dade and Fairfax County continued to innovate, developing and testing new approaches to rescue that would prove to be beneficial when the U.S. government embarked on a campaign to establish a nationwide system of USAR teams. In 2003, USAID/OFDA included the LACoFD in its system of internationally-deployable USAR task forces.

Based on a mandate from Congress in 1990 (after the Loma Prieta quake and Hurricane Hugo), FEMA developed 25 operationally functional, 60-person, multi-discipline USAR task forces in just two years. The development marks one of the most ambitious and successful programs ever produced by the federal government. Rarely has a government program achieved its stated goals with such clarity of vision and purpose. And rarely has there been such a successful partnership between the federal and state governments

and local fire/rescue agencies and civilian experts. FEMA was handed a mandate to develop a national USAR system for disasters, and the nation's fire/rescue services jumped in with both feet, making the program happen with such speed and at such a cost value to the federal government that it has rarely been equaled.

Since 1992, when the first FEMA USAR task forces were deployed to Hawaii during the nation's response to Hurricane Iniki, these federal resources have vastly improved the heavy/technical rescue capabilities of the U.S. government. The successful implementation and operation of FEMA's USAR task force system proved to be a watershed event in terms of timely and expert government-sponsored response to events involving people trapped in collapsed buildings and other predicaments. FEMA USAR task forces were established when Hurricane Iniki, the Northridge earthquake, the Oklahoma City bombing, and other major disasters challenged the nation's fire/rescue services in rapid succession. FEMA's USAR task forces (now numbering 28, staffed by 68 rescuers per task force) have demonstrated that the timely infusion of well-equipped, highly trained, experienced, and properly organized teams is of great value to local ICs faced with large-scale, complex, and personnel-intensive SAR problems.

In the wake of these events, it's clear that the USAR task forces have revolutionized the way in which many large-scale SAR disasters are managed. USAR task forces are designated as local resources for local disasters, statewide resources for incidents that rise to the level of state disasters, and as national resources for federally-declared disasters. They are arguably the most unique tactical resource to happen in emergency management in the history of the United States.

On the west coast, the effectiveness of the eight FEMA USAR task forces based in California was extended in 1995 when the governor's OES mandated that each task force be prepared to deploy fully-trained, 15-person swift-water rescue task force components anywhere in the state. Equipped with state-purchased inflatable rescue boats (IRBs), personal watercraft (PWC), PPE, and other state-of-the-art gear, these OES/FEMA swift-water rescue task forces were pressed into service during major floods that struck northern California for three weeks in 1995 and 1996. Whenever a California-based FEMA USAR task force is dispatched to a hurricane or other disaster where flood rescue is a potential problem, they are equipped and trained to deal with any swift-water rescue hazard. Other FEMA USAR task forces, including those based in Pennsylvania, Arizona, and Texas, have expanded their capabilities to include flood and swift-water rescue operations.

Today there is a growing recognition that most hurricane fatalities are caused by flood events, and that unusual events like volcanic eruptions and earthquakes may be accompanied by massive flooding. In fact, moving water accounts for a large portion of the nation's annual disaster-related deaths. This is prompting the federal government to embrace the concept of a timely response of expert swift-water rescue capabilities (including swift-water-capable FEMA USAR task forces) to support the efforts of local fire/rescue agencies during flood-related disasters. There is a move afoot to mandate and fund FEMA to expand the mission of all USAR task forces to include flood and swift-water rescue operations. The FEMA USAR task forces are clearly the most logical, cost-effective, and timely resource to apply to flood rescue operations that exceed the capabilities of local and state agencies. California and Pennsylvania have already proven the usefulness of this approach because swift-water rescue-trained members of USAR task forces from those states have already saved lives during flood disasters. As of this writing, Congress is considering this issue, including sufficient funding to make the program a success.

All FEMA USAR task forces are also being trained and equipped to conduct SAR operations in environments contaminated by weapons of mass destruction (WMD) events, including those involving nuclear, chemical, biological, explosive, or incendiary devices and hazards. They are rapidly becoming the principal support resources for local fire/rescue departments in the event of WMD attacks. And with the establishment of the U.S. Department of Homeland Security, which now controls FEMA and a host of other federal response agencies, the new paradigm in rescue is even more apparent.

These and other developments represent a sea-change in disaster response theories and practices nationwide. These stem, in part, from a similarly significant change within the nation's fire service, which is clearly among the most critical components of timely disaster-related SAR. These developments aren't limited to disasters. These resources provide timely expert response to technical SAR emergencies and disasters on a scale not even dreamed of just 20 years ago.

New Legal Risks for Firefighters and Other Rescuers

Perhaps a story from another part of the world best exemplifies the importance of avoiding jurisdictional squabbles, as well as illuminating emerging legal risks for rescuers who fail to act in the public's interest. Consider an unfortunate incident in the Taiwanese province of Chiayi. In July 2000, four citizens there perished in a raging flood after a horribly botched, hours-long rescue attempt during a raging flood on the Pachang River. As of this writing, a police helicopter pilot faces a ten-year prison term and several firefighters are threatened with prison time ranging from five to ten years, following their indictment on charges that include breach of duty, professional negligence, and manslaughter, for their roles in the failed rescue operation.

The victims had become trapped on the far side of a river suddenly swollen by flash flooding that swept the region. As Taiwanese news cameras televised the drama live across the nation, the four people clung to vegetation, buffeted by the torrent. Firefighters were unable to get rescue lines across the wide channel, and the situation became increasingly desperate. It was clearly a situation that called for the use of rescue helicopters. But there was a dispute over which agency should attempt the rescue. But in this case, it wasn't a race to see who could do the rescue. This time, the local police pilot argued that it wasn't his responsibility, and several of the firefighters also delayed efforts to initiate an effective rescue.

An argument ensued with the military over who should risk lives to attempt the rescue. Meanwhile, the victims clung to precarious perches for several hours, just as the debacle was being aired live on national television news programs. Finally all four victims were swept away by flood waters before a successful helicopter-based rescue attempt could be mounted. The problems associated with this unsuccessful operation drew the ire of angry Taiwanese, who wanted blood in response to the obvious disregard for innocent human lives that was demonstrated by the on-scene officials.

After the helicopter pilot and firefighters were indicted, a Taiwanese chief prosecutor was later quoted as saying, "It was (the firefighters' and pilot's) duty to provide disaster relief, but they were unwilling to carry it out and passed it on to other people." As a result of their failure to act in the defense of innocent victims, the pilot and firefighters may spend up to a decade in prison. Across much of Taiwan, the decision was hailed as a victory by citizens who were outraged that no one took control of the incident to lead the rescue efforts from the early stages.

In Taiwan, as in other places in the world, there is an implied expectation that firefighters and other officially sanctioned rescuers will take calculated risks to save the lives of trapped or missing people during the course of an emergency. In some places, the fire department or other public safety agency is specifically designated the agency with primary responsibility for conducting rescue operations.

Could American firefighters and rescuers one day be prosecuted criminally for refusing to initiate rescue operations under similar conditions? Some legal analysts maintain that firefighters and other rescuers become responsible for the welfare of victims only after making physical contact with them, or after actively attempting to reach them. But other experts contend that it's a matter of time before a criminal case is brought against rescuers who refuse to initiate life-saving operations, or who bungle the rescue because they lack a reasonable level of preparedness for well-known local hazards. With the recent adoption of national standards like NFPA 1670, NFPA 1006, and other rescue-related standards, there are few excuses for modern fire/rescue agencies to remain unprepared for such hazards.

Differences and Similarities in East and West Rescue Practices

When considering rescue-related tactics, strategies, philosophies, capabilities, and other factors, there's a natural tendency to compare the east and west coasts and the Midwest states of the United States (just as regional comparisons are made regarding art, music, education, and other elements of our vast and multifaceted societies). And while it's fair to acknowledge certain differences in the way rescue is accomplished in different parts of North America, perhaps more striking are the *similarities* between East, Midwest, and West when it comes to the basics. For example, Canada's approach to rescue is consistent with that of many of her southern counterparts, and the same can be said in some parts of Mexico. In fact, it can be said that the fire/rescue services of Europe, the former Soviet Union, Australia, South Africa, South America, the Caribbean, and Asia share important consistencies with North America when it comes to rescuing lost or trapped people. And it's true to say that in rescue, as in disaster response, the line of difference between nations is becoming increasingly blurred.

Still, it's instructive to look at the prominent differences that do exist between the ways in which rescue is viewed by fire/rescue agency administrations (and hence how it's managed) from east to west in the United States. In some respects, the genesis of rescue differs from coast to coast, due in part to marked differences in terrain and climate. The prominent effects of seismic activity (and seismology's sister hazards of structural collapse, conflagration, mass casualties, tsunamis, and landslides) inherent to the west coast are a definitive factor in the development of USAR as a formal discipline. Another contributing factor is the effect of severe weather in the West, which in combination with high mountains, steep terrain, and the after-effects of huge wildland fires (i.e. massive floods, mudslides, and mud/debris flows), tends to create distinctive hazards. The Midwest, with its tornadoes, severe thunderstorms, and accompanying flash floods, seems to be the target of weather-related destruction on a more consistent annual basis than either of the two coasts. And the East, with its older buildings, its severe winters, and the annual specter of hurricanes, is the site of its own share of rescue problems that generally aren't found in the west or Midwest. As a result, rescue has traditional differences that are regional in nature.

On the east coast and in many parts of the Midwest, dedicated fire department rescue companies and heavy rescues have been in place for decades. It's simply been assumed that these units will be purchased, equipped, and staffed as part of a fire department's standard regimen. In many of these places, heavy rescue

and collapse rescue have long been considered a priority by fire department administrators, partly because of the history of firefighters being buried by collapsing buildings during fires. Concern for firefighter rescue is in fact closely related to the genesis of the FDNY's pioneering rescue companies. Heavy rescue capabilities have long been a priority in the Midwest and along the east coast. Today, fire departments on the east coast remain on the vanguard of topics like rapid intervention and firefighter self-rescue.

Although the West has generally picked up on the success of this approach to rescue and rapid intervention, many western fire/rescue agencies still have much to learn from the East and Midwest in this respect. The newfound nationwide emphasis on urban search and technical rescue has simply strengthened existing rescue programs in many midwestern and eastern fire/rescue agencies, whereas in many parts of the West, the formal discipline of rescue is treated as a relatively new phenomenon (some would say a novelty). It's ironic to note that in many parts of the disaster-prone West, rescue as a fundamental role of the fire service deserving of dedicated staffing and specialized apparatus is a relatively new concept, having evolved only during the past decade or two (with some notable exceptions). (Fig. I–6)

Fig. I–6 Modern USAR Companies Carry Tools Unheard of 10 Years Ago

The prevailing view among many western fire departments has traditionally been that firefighters, by virtue of their fireground/EMS skills and good old-fashioned common sense (not to mention trade skills), are essentially jacks of all trades who will always find some way to handle rescue predicaments. This may change somewhat as firefighters coming through the ranks tend to be more technology-oriented, but it still holds true to some extent. Plus, cities in the West, being younger, don't experience structural collapse with the frequency of older cities in the East (with notable exceptions like Los Angeles and San Francisco, which are periodically shaken by damaging earthquakes and other natural and man-made disasters).

But even in the face of strong evidence for the need to develop specialized rescue units, the generalist approach was emphasized to the point that rescue training and dedicated rescue companies were for decades considered unnecessary. It was assumed that in the absence of formal rescue training and equip-ment, firefighters would simply figure out some way to locate and extract trapped people, even in highly technical environments like cliff rescue or in collapsed buildings in the aftermath of an earthquake. Until the 1980s, standard rescue training throughout much of the western U.S. fire service was generally limited to vehicle extrication and basic rappelling with ladder belts on manila rope. And formal training for topics like structure collapse, trench/excavation collapse, high-angle rescue, swift-water rescue, and other forms of rescue, was practically nonexistent.

Something was needed to break the stalemate between those who considered rescue a *sideline* of the fire service and those who understood rescue as a *fundamental role* of fire departments. Fortunately, lessons were learned and heeded by western fire/rescue leaders, prompted in part by disasters and other prominent rescue-related events. For example, the 1971 Sylmar earthquake prompted the California state fire marshal to commission a group of fire service experts to develop a course to enhance the disaster rescue skills of firefighters. Thus was born the seminal course known in those days as Heavy Rescue Systems I (the forerunner of Rescue Systems I and II, which have been adopted as national courses and eventually evolved into courses like Structure Collapse Technician and FEMA USAR Rescue Specialist). High-angle rescue was also introduced as a formal discipline for firefighters.

The deadly El Niño floods of 1982 and 1983 prompted fire departments in San Diego, Orange, and Los Angeles Counties to institute formal swift-water rescue systems. Then the Whittier earthquake that struck Los Angeles County in 1987 and the Loma Prieta earthquake that hit San Francisco and Oakland in 1989 made it clear that advanced collapse rescue units were needed by local fire departments faced with dangerous seismic risks. Soon thereafter, dedicated rescue companies and task forces in fire departments sprang up along the west coast in the late 1980s and through the 1990s.

Which leads us to an intrinsic truth. When tragic and world-shaking events like the 9-11 attacks over-whelm local, regional, and state capabilities, it is now possible for rescue teams from all corners of the United States to come together as one large team to conduct coordinated, round-the-clock, long-term, urban search and technical rescue operations. Urban search and technical rescue was once treated as an exotic afterthought in some realms of the fire service. In fact, the strongest proponents of fire department-based rescue were once considered unorthodox eccentrics, or even zealots. Today though, fire department-based rescue is accepted practice across North America. In short, urban search and technical rescue is now *one of the things that firefighters do.*

Making a Difference

In the not-too-distant past, the odds were decidedly stacked *against* the rescuer, but increasingly, the odds are now being stacked in *favor* of rescuers through better training, equipment, and planning, and the availability of advanced rescue resources. Armed with better equipment, specialized training under realistic conditions with highly developed plans and systems, backed up by highly capable rescue resources, today's firefighters are far better prepared to manage the consequences of high-risk, complicated rescue incidents.

Many existing rescue systems have been reinvigorated by the newfound emphasis on USAR and disaster rescue. In short, it's been nothing short of a revolution. One major reason for this revolution is the rising recognition of SAR (in all its forms) as a fundamental role of the fire service. Another is the public's perception of the fire department as the solution not only for fires, but for practically all non-law enforcement emergencies. When they dial 911, citizens generally assume that fire engines and other apparatus will show up, generally within a few minutes, with highly-trained firefighters prepared to render effective aid in almost any circumstance. This perception has been reinforced in recent years by the daily performance of the ever-more-proficient fire service, as well as a string of successful, high-profile rescues under conditions that might have previously denied success to less-prepared fire/rescue agencies.

Evolution and Tradition

The fire and rescue services are creatures of evolution, however painfully slow that progress might seem to be at times. Our chosen profession (like so many others) is sometimes plagued with a form of institutional amnesia. Many of us tend to forget that modern fire and rescue service practice derives (in part) from knowledge and traditions inherited from long-deceased predecessors, knowledge and experience that's been hammered into contemporary form to suit the needs of modern fire and rescue culture.

Although we are the benefactors of modern technology applied to fire and rescue work, it's also true that many modern rescue and firefighting concepts are actually rooted in principles developed by firefighting and military forces that existed long before the time of Christ, Mohammed, and Abraham. Others are the echoes of tactics employed by fire brigades that protected European, Greek, Middle Eastern, and Far Eastern civilizations thousands of years ago. Even some aspects of modern fire department organizational structure and management are rooted in ancient fire brigade models like the *Militia Vigilium*.

Consequently, evolution and tradition are constant factors in shaping the modern fire and rescue services. Over millennia, some of the more helpful principles have been selected and passed on by long-dead fire and rescue authorities in the form of traditions, sometimes supported by written lessons. Succeeding generations have built upon the existing body of knowledge that was their heritage, and, in turn, added their own interpretations and knowledge based on their more modern experiences. If not for this evolutionary process, each succeeding generation of firefighters and rescuers would be hamstrung by the need to learn the basics from experience alone.

This book series is a small and humble attempt to support the evolutionary march of progress with regard to rescue by presenting a certain viewpoint on assessing, planning, and managing urban search and technical rescues operations...which hopefully will stimulate conversation and thought about the topics herein. While some of the information may have a certain *West Coast* flavor (owing to the experiences of the author and his colleagues), it's also been deeply influenced by the shared experiences of members of active rescue companies and USAR units across North America, from New York to Seattle, from Albuquerque to British Columbia, from Florida to Phoenix, from L.A. to

Oklahoma City. This book is also based on the experiences of fire and rescue agencies in places as far-flung as Australia, Asia, Africa, Europe, South America, the Mideast, and the Caribbean. Clearly, the fire and rescue services are a truly global calling, and we continue to have much to learn from one another, regardless of where we come from.

It should also be noted that some of the most important information herein came from firefighters and rescuers who make a daily practice of asking the question: *What if?* Without intelligent people asking that question about the potential for unexpected events, the fire and rescue services would be left with lessons gained strictly through experience, much of which comes at a painful cost. It's important to keep asking *What if?* because the world is changing fast and potential rescue problems—many of which would never have been contemplated just a few years ago—are developing with equal speed.

In those terms, this book is also intended to spur the search for answers to the question: *What if?* The ultimate desired result is to improve the speed, effectiveness, and safety with which firefighters and other rescuers locate and extract trapped victims to help accomplish the main goal of protection of life. By practically any measure, both daily rescues and disaster SAR operations in North America are now conducted faster, more efficiently, and more safely than ever before. These revolutionary improvements aren't by any means limited to North America. In many parts of the world, fire departments and rescue teams safely conclude SAR operations that would have been unthinkable just two decades ago. And revolutionary changes continue to occur at an ever-more rapid pace as new technologies, funding sources, and other resources are directed at improving SAR capabilities.

Yet cutting-edge urban search and technical rescue capabilities remain an elusive goal for many fire/rescue agencies that constantly struggle to balance the need to properly train and equip members to manage emergencies against the desire to develop effective rescue response. In too many places, the ability to implement a formal rescue or USAR program is considered a luxury, something to be done after legal mandates and other local priorities are addressed. It's not uncommon to find that rescue is treated as an afterthought in terms of dedicated staffing, training and continuing education, budgeting for equipment, and development of operational guidelines. Who suffers worst from that dynamic? In some cases, it's an innocent victim awaiting help. In other cases, it's the firefighters and rescuers themselves who may suffer for lack of advanced rescue and rapid intervention capabilities.

Add to that the difficulty encountered by firefighters who lack the benefit of routine exposure to technical SAR emergencies, mostly because true *working* rescue emergencies are relatively rare in many regions and because realistic training is often hard to come by. When it comes to disaster SAR, it's even more difficult for local fire departments to develop the levels of experience necessary to effectively manage the multiple demands. Consequently, major rescues and disasters are sometimes marked by the misappropriation of emergency resources, mistaken priorities, and questionable strategies implemented by decision-makers lacking solid rescue experience. Other rescue events are plagued by excessively dangerous and time-consuming tactics by inexperienced, ill-trained, and improperly equipped rescuers. Sometimes, making the decision to request outside assistance, even when the local jurisdictional agency is clearly overwhelmed by the scope or nature of the incident, caused unnecessary and lengthy delays.

Yet despite these difficulties—which are often regional in nature—the following statements are clearly true. Never before in history has the fire service been better prepared to handle the wide range of technical SAR emergencies that occur every day. The world has never been better prepared to manage disaster SAR operations. In short, many challenges remain for those charged with managing and conducting technical SAR operations, including firefighters assigned to engines, truck companies, rescue and USAR companies, paramedic squads, and other fire department units. Equal challenges remain for personnel assigned to local and regional rescue teams, as well as state, national, and international USAR teams. But two decades of renewed emphasis in SAR has invigorated funding and other support for improvement of these capabilities. For the tailboard firefighter, the rescue company member, and the chief officer, this equates to a time of discovery and improvement in the field of SAR.

For those whose agencies aren't yet committed to developing rescue and USAR programs, this book offers realistic and field-tested guidelines for establishing viable response systems. For those whose agencies aren't capable of providing the most advanced levels of training, this may provide a sort of guide for developing and implementing effective training regimens. And for firefighters who don't benefit from daily exposure to rescue-related emergencies, this book is intended to convey valuable lessons gleaned from other firefighters whose daily role is to manage technical SAR operations.

Rescue in its modern form is in fact a diverse collection of disciplines, often related by common equipment, ruled by common methodology, and conflicted by different regional names and rules. Despite ever-greater attempts at national and international standardization by organizations such as the NFPA, the National Association for Search and Rescue (NASAR), several state offices of emergency services, and the OSHA, and the INSARAG, the vast majority of fire/rescue agencies still operate with wildly varying standards for training, equipment, and continuing education. Hopefully this situation will continue to improve in the coming years.

1

Planning for "Daily" Urban Search and Technical Rescues

Why Plan For Rescue?

Rescue is among the most basic jobs in public safety. Few roles are more central to the work of firefighters, rescue teams, and others charged with protecting the lives of citizens. With each passing year, the task of rescue becomes more embedded in the mission of the fire service, even in regions where technical rescue was traditionally relegated to volunteer SAR teams or law enforcement agencies. Today, the average citizen assumes their local fire department is prepared to extract trapped victims from practically any situation; or at least to effectively evaluate the situation, determine needs, and request the appropriate resources to get the job done. This is a reasonable expectation in the modern world, considering the proven record of the fire and rescue services in freeing people who might otherwise have little hope of safe rescue. Success breeds success, just as it breeds expectations of continued triumph.

Accordingly, planning and preparation for effective management of daily rescues is an equally basic respon-

sibility of firefighters, fire department officers, SAR team members, and others charged with the resolution of rescue emergencies.

Rescue as risk management

Broken down to its essence, rescue is a basic form of risk management. As firefighters and rescuers, we help society manage the potential risk of fires, medical emergencies, transportation accidents, industrial mishaps, recreation accidents, and other daily emergencies. We manage the risk posed to people who—for limitless reasons—become trapped and require physical rescue. Clearly, then, rescue is one of the most basic forms of consequence management, one that is applied during and/or after a mishap with trapped or missing victims.

Although they have limitless causes, physical entrapment situations are often predictable or even preventable. Predictability and preventability are principal concepts of risk management stressed by Gordon Graham, respected attorney and consultant to fire and law enforcement agencies across the United States. Indeed, history conclusively supports Graham's case: from the collapse of double overhead walkways in a crowded Kansas City hotel in the 1980s to construction workers trapped in the collapse of non-shored trenches; from children being swept away in flood-control channels in urban areas to workers trapped in machinery; from window washers trapped on high-rise buildings to the sudden collapse of burning buildings. In almost every case, post-accident (or post-failure) analysis demonstrates a chain of events that might have been anticipated or prevented with more effective planning, more aggressive prevention measures, better situational awareness, and more time.

Obviously, it's easier with hindsight to determine the cause of events that have already occurred. However, with proper attention to planning and a pro-active stance, it's sometimes possible to foresee events that can be prevented, or (if they can't be prevented) whose consequences can be managed or limited.

Prevention does have its limits. Human nature, our physical environment, and fate being what they are, people seem compelled to tempt fate even in the face of clear warning signs. As a consequence, preventable accidents and adverse events will continue to occur. Man-made structures and machines will continue to fail catastrophically. People will fall victim to natural disasters. People will make mistakes and suffer errors of judgment. They will venture off trails, fall off cliffs, and crash their cars over the side of mountain roads. And, in the modern world, people will continue to be trapped or end up missing as a consequence of terrorism, arson, and civil unrest.

In short, prevention and crisis management don't always work, and therefore people will continue to find themselves trapped in unusual predicaments that require physical rescue by fire/rescue professionals like you. This means that you and other rescuers must be prepared to manage the risks of a practically unlimited variety of urban search and technical rescue emergencies.

Determining responsibility

The first step in the planning process for daily rescue operations is to clearly identify which agencies have formal responsibility for rescue operations and to charge these agencies with the mission to conduct realistic and comprehensive planning. In some places, there is only one rescue agency to handle daily emergencies, and there is no question about who is in charge or responsible. In other places, there may be more than one agency with rescue capabilities and responsibilities, and at least a gentleman's agreement on who is primarily responsible. In some places, there is a clear sharing of rescue responsibility. In still other regions, there may be multiple rescue-capable agencies whose members actually compete to arrive on the scene first in order to take control of the scene and reach the victim first. Obviously, this last arrangement can lend itself to a number of public safety complications and legal entanglements.

Optimally there will be some written policy, law, or protocol that spells out responsibility for the broad range of rescue operations that may be required in a given geopolitical area. It may be in the form of a state, county, or city charter; in a written policy or protocol; as part of an inter-agency agreement; or some other way that clearly delineates who is responsible for rescue, which agencies are to be notified for emergency response to rescue emergencies, which agency is the primary responsible entity, and which agencies are secondary or support responders, and how incident command is to be handled.

It's important for rescue authorities to determine these factors and to ensure that they are clearly spelled out because they will form the basis upon which many decisions about planning, preparedness, training, and response will be made.

Evaluating Local Hazards

Naturally, the first step in rescue planning is a realistic evaluation of local conditions and hazards. By deducing the range of daily rescue situations likely to occur locally, fire/rescue officials can extrapolate to develop an assessment of the likely rescue emergencies and then match their urban search and technical rescue capabilities to the risks. Achieving a high level of predictability based on local conditions and demographics has proven to be a potent advantage for any organization tasked with managing rescue operations.

Properly employed, accurate risk information can help guide fire/rescue officials in the plans, systems, and resources capable of properly managing the consequences of mishaps in industry, at home, on roads and highways, at sea, in the mountains, and practically anywhere else where daily rescues occur. By recognizing the benefits of properly assessing the potential for daily urban search and technical rescue emergencies, and by supporting efforts to quantify them, emergency officials will be ahead of the proverbial power curve in terms of preparing their respective agencies to manage the rescue emergencies likely to be encountered.

Naturally, a number of factors and complications challenge the success of rescue operations. Nevertheless, rescue can often be safely accomplished if emergency responders and fire/rescue authorities are adequately trained and experienced to recognize local hazards that translate to rescue emergencies. In contrast with some types of disasters whose specific effects may be difficult to predict with sufficient detail to anticipate every eventuality, the effects of daily rescues can often be anticipated with a reasonable degree of accuracy.

For firefighters and other emergency responders to conduct a realistic evaluation of daily rescue hazards, it is important first to develop an intimate understanding of the local conditions, to recognize conditions that typically lead to rescue emergencies, to understand the dynamics of typical rescue emergencies, and to be alert for signs of trouble that aren't obvious to the untrained eye. In principle, this process is akin to a fire-ground pre-attack plan. It can also be likened to a size-up of a rescue scene, except this one is performed *before* the rescue occurs.

Figure 1–1 shows a daily rescue emergency situation. What is the potential for a daily rescue emergency like this in your district resulting from daily causes like gas explosions, vehicles crashing into buildings, or even terrorist activities? How ready are your personnel to assess the scene, determine the need for structural stabilization, place shoring in a timely manner to support SAR, and to locate and extract victims?

Fig. 1–1 Daily Rescue Emergency Situation

A variety of approaches may be employed to assess local rescue hazards. Some agencies choose to send hand-selected groups of personnel to formal rescue training courses. Another option is to develop exchange programs with agencies that respond to similar hazards to expose personnel to actual rescue operations and to observe how they are mitigated by outside agencies. Personnel who attend formal training or participate in exchange programs, in turn, may be assigned to committees, work groups, or other assignments where they are responsible for conducting local hazard studies and developing plans based on their training and experiences.

Some agencies choose to hire outside consultants to study local hazards and make recommendations for mitigation. Agencies with the ability to staff dedicated rescue companies or USAR units have a natural advantage because these units are ideal resources for conducting local rescue hazard assessments. However, it's not necessary to have dedicated rescue units in order to be prepared for rescue. Certainly there are a number of options for modern fire/rescue agencies. Successful alternatives to dedicated rescue companies have included the use of existing ladder (truck) companies and even engine companies as dual-mission units that specialize in urban search and technical rescue. Some departments designate entire fire stations as rescue companies, USAR companies, or rescue units, and give them the mission of training, assessing, planning, and responding to rescue emergencies. Other agencies send all their firefighters through formal rescue training, knowing that any of them may be faced with a rescue emergency at any time. Still other agencies use combinations of these options.

In any case, rescue hazard evaluation criteria should include not only the physical environment (e.g., terrain, hydrology, seismology, roads and freeways, and buildings) and demographics but also a study of weather patterns and other interpretive factors. The patterns that emerge may identify geographic locations where trouble can be expected under certain conditions (providing an opportunity for mitigation).

For fire/rescue agencies that simply lack the financial or other resources to provide specialized training for firefighters and other personnel—and for those without the wherewithal to hire consultants—it is still possible through the use of generic guidelines or instructional materials to prepare first responders to recognize basic indicators of daily rescue problems.[1]

In combination with an historical review of past incidents based on official records such as 911 telephone system data, fire/rescue dispatch and response data (an especially potent source of information with the advent of computer-assisted dispatching and record-keeping), the National Fire Incident Reporting System (NFIRS), and other official records, this approach may prove effective in recognizing and assessing local hazards. Before he lost his life in the collapse of the World Trade Center towers during the 9-11 attacks, Ray Downey, an esteemed deputy chief of the FDNY's Special Operations Command, put it simply: "Review past incident histories. Know which kinds of incidents are most common and prepare for them. But always expect the unexpected."

Lacking a reliable, long-term record of local rescue incidents, some agencies may find it helpful to consult other sources for anecdotal data. These *unofficial* sources of information about local rescue patterns and trends may include police and fire department dispatchers, the local coroner's office, newspapers and other news media, and field personnel (active *and* retired), as well as others with a broad sense of the jurisdiction's history. It may also be beneficial to check with outside agencies such as public works, flood-control districts, and other agencies that deal directly or indirectly with public safety. Significant patterns may even emerge from a file search of local newspapers, obituaries, and other unconventional sources.

If historic patterns of daily rescues are indicated from official and/or unofficial records, they can often be found to correlate with certain terrain features, including mountains, coastal cliffs, rivers, mud flats, lakes, and avalanche-prone slopes. Correlations can also exist with certain weather patterns or microclimates; with man-made hazards, such as Arizona crossings,[2] low head dams, industrial complexes, amusement parks, railways, construction sites, tunnels, local and regional aircraft patterns, freeways, man-made confined spaces, homes and non-residential buildings, and with other potential rescue hazards. Recognition of these and other daily rescue hazards enables fire/rescue professionals to develop appropriate plans, to mandate appropriate training, to acquire the necessary equipment, and to develop an emergency response system equal to the problem.

Development, maintenance, and use of an accurate database will assist in identifying the causes of rescue emergencies and should be addressed in the planning and emergency response processes. Perhaps the simplest and most effective means of recognizing and assessing daily rescue hazards is to assign fire/rescue personnel to survey their jurisdictional areas with their own eyes. Armed with a basic list of parameters and basic training in hazard recognition, even non-rescue-trained fire/rescue personnel can begin charting the types and locations of various rescue hazards. At the least, this may establish the need for additional training and/or consultation to address hazards as they are discovered and documented.

Examples of daily rescue hazards

In simple terms, if any of the following hazards are found within (or near) an agency's jurisdiction, there is a potential daily rescue problem that should be addressed by plans for effective management when it occurs:

- Natural rivers or streams, whether flowing year-round or seasonal only

- Mountains

- Industrial complexes

- Potential terrorism targets, such as embassies, churches, mosques, and government buildings

- Flood-control channels

- Refineries

- Water treatment plants

- Airports

- Mines and tunnels

- Rock quarries

- The ocean

- Coastal cliffs and bluffs

- Wildland fires in steep terrain (indicating the potential for massive post-fire flooding and mud and debris flows, as well as increased runoff in urban areas)

- Dams

- Arizona crossings

- Construction projects

- Railroads

- Regions subject to flash floods, rockslides, or mud and debris flows

- Subway or metro-rail systems

- Amusement parks

In Figure 1–2, an L.A. City Fire Department USAR firefighter downclimbs past emergency shoring during rescue operations to uncover a construction worker who was buried beneath 27 ft of soil when the walls of a bore hole collapsed and pulled him into the auger. Note the tag line attached to the firefighter's harness and the use of shoring to support the unstable walls of the round hole with its mixed running and clay soils. The black tube is a commercial-grade hydro-vac hose, which is being used to expedite the removal of dirt from atop the victim in this confined space.

Fig. 1–2 USAR Firefighter Involved in Rescue Operation

Construction sites are commonly associated with daily rescues like this one, in which L.A. City, L.A. County, and Vernon fire department personnel worked under a unified command to locate and extract the buried worker, who did not survive his entrapment. Daily USAR emergencies encompass a wide range of mishaps, including:

- Trench and excavation collapse

- Structure collapse

- Vehicles over-the-side (mountain roads, freeway overpasses, etc.)

- Cliff rescues

- High-rise emergencies (fires, stranded window washers, etc.)

- Airplane and helicopter crashes

- Unconventional aircraft mishaps (hot air balloons, hang gliders, etc.)

- Train derailments and collisions

- Shipping and boating accidents

- Large vehicle extrications (semi trucks, bulldozers, cranes, etc.)

- Industrial accidents (machinery entrapment, explosions, etc.)

- Confined-space rescue

- Tunnel fires, collapses, flooding, and other emergencies

- River and flood rescues

- Jumpers on bridges, towers, and high-rise buildings, etc.

- Mud and debris flows

- Landslides

- Explosions

- *Maydays* and *firefighter down* situations on the fireground

- Myriad other situations that leave victims lost and/or trapped

It can be said that where disasters are common there is a higher than normal potential for the occurrence of daily technical rescue emergencies as well. Equating the presence of volcanoes, wildland fires, earthquake faults, landslides, floods, and other disasters with daily rescue hazards may seem preposterous at first, but history tells us that conditions that lead to disasters are also prone to cause daily rescue problems.

Unusual rescues

An oft-heard comment in public safety is: "We've never had an incident like that yet; so why should I be concerned about it now?" In some places, these are the equivalent of the proverbial famous last words. Few regions in the world are exempt from technical rescue hazards. Just as being paranoid doesn't ensure that someone isn't really following you, the absence of experience with a certain type of rescue in your jurisdiction doesn't mean that it won't happen tomorrow, next week, or next year.

In Figure 1–3, L.A. County USAR firefighters and an industrial fire brigade member at the U.S. Borax Company (in neighboring Kern County, California) attempt to uncover a worker buried to his chest in Borax inside a 100-ft high silo. The man's partner was buried below him and appeared to have been alive for nearly $1^1/_2$ hrs in an air pocket (based on the first victim's report that his colleague was grasping at his ankles from below). Working with the Kern County Fire Department (which requested the L.A. County USAR unit for mutual aid), the firefighters managed to rescue the first man alive, but by the time they were able to tunnel down to unbury his partner, he was deceased. Workplace rescues like this confront firefighters across the industrialized nations. Does your department respond to places where daily industrial rescues like this can happen? Are your department's personnel prepared to conduct the rescue in a reasonably safe and timely manner? Are your incident commanders prepared to request mutual aid in a timely manner when extra assistance is needed? Does your dispatch center maintain an up-to-date database of specialized resources that may be required to manage unusual rescue operations?

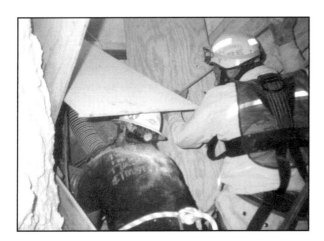

Fig. 1–3 Attempt to Rescue a Worker Buried in a Silo

It may not be possible to be completely ready for every possible type of rescue that can happen in a given jurisdiction, and clearly there's a need to balance the *risk vs. gain* equation in a way that ensures adequate emergency service for the public and a reasonable degree of safety to fire department personnel without breaking the bank. Here, then, is a good rule of thumb: if there are identifiable rescue hazards in your jurisdiction, there's a real potential for daily rescue emergencies, and therefore a need for effective rescue resources to address these situations when they occur.

The central problem, of course, is recognizing and assessing the threat with an acceptable degree of accuracy and determining to what extent any particular hazard requires action to prevent loss of life. What follows is a sample of rescue hazards, with emphasis on assessing the potential for daily rescue emergencies.

Assessing swift-water and flood rescue hazards

When cities and towns are bisected or bordered by natural waterways, the potential for rescue situations is rather obvious, even if they are merely seasonal streams that flood only during major downpours. However obvious the hazard appears, the danger may not be fully appreciated by the untrained eye. In the case of swift-water rescue, the danger may be less apparent (and thus more difficult to assess) in remote rivers and streams.

Assessing the severity of swift-water and flood rescue hazard is somewhat dependent on the observer's ability to *read* the waterway and recognize the consequent danger signs: essentially understanding the manner in which moving water reacts to changes in river courses, differences in gradients, and obstacles like trees and boulders. Experienced swift-water rescuers, whitewater rafting guides, kayakers, and others who spend a lot of time on rivers, are accustomed to *reading* the flow of water in order to avoid being subjected to highly dangerous hydraulics and currents and to choose the safest routes through rapids. Firefighters and other first responders can be trained to read a river in much the same way, with an eye for potential rescue problems.

Mountain and wilderness rescue hazards

Mountain and wilderness rescue are among the most common forms of technical rescue in many areas. These rescues are certainly among the most challenging and hazardous, occurring as they do in remote areas where there may be long response times, drastic weather conditions, poor communications, difficult access, troublesome logistics, delayed backup from other units, unusual predicaments, and other complications. The first steps in preparing for these rescues are determining the level of responsibility for performing them and assessing the local hazards.

While the jurisdiction of some fire departments is restricted within the borders of a city or town, other fire departments have responsibility for providing fire and rescue service in wilderness areas. The presence of mountains in the jurisdictional area indicates the potential for an entirely new range of daily technical rescue emergencies that can occur far from the borders of any town or city. With the expansion of formal rescue systems across the fire service, it's increasingly common to find local fire departments responding deep into the mountains and other wilderness areas to conduct high-angle rescue and other SAR operations.

In some states and counties, law enforcement may have primary responsibility for the ultimate resolution of mountainous rescues (and hopefully for supporting the fire department's primary role of providing pre-

hospital care for injured victims in wilderness areas). In some places, volunteer SAR teams operate under the umbrella of local law enforcement agencies or a county or state charter, and they may respond with or without the fire department (depending on local protocol, charters, and legal responsibility).

Frequently it's the fire department's primary responsibility to provide pre-hospital care of injured victims, regardless of who has overall responsibility to conduct SAR in wilderness areas. This requires firefighters and EMS personnel to use any appropriate means (i.e., ropes, helicopter hoists, etc.) to make physical contact with trapped or stranded victims as soon as feasible to assess, stabilize, treat, and remove injured parties from harm without delay.

In some instances, the local fire department may have primary responsibility for the safe and effective resolution of all SAR operations in mountainous areas. In fact, the fire department may be the only local agency with formal SAR capabilities. In these cases, local law enforcement agencies may be responsible for assisting with such tasks as traffic control, handling deceased victims, and conducting accident investigations, while the fire department handles the actual search and technical rescue operations.

In some areas, the federal government may be the primary SAR agency. This includes national parks and forests, navigable waterways, the coastline, and certain other places. Military resources—especially helicopter rescue teams—may be employed during the course of mountain SAR operations. In places like Denali National Park (Alaska), the military (in the form of helicopter-based U.S. Air Force para-rescue jumpers and fixed-wing aircraft units) plays a central role in mountain SAR.

Certainly, as one travels across the United States and other industrialized nations, there are other combinations of resources—and responsibilities—for mountain SAR operations.

A simple rule of thumb is: if the jurisdiction of your department includes mountains and wilderness areas and if people live or visit those areas (even if they fly over them in private or commercial aircraft, or boat in them), there is a need for mountain or wilderness rescue capabilities. If the mountain and wilderness areas receive snow, there is a potential need for ice and snow rescue capabilities, which may include avalanche rescue, ice scuba rescue diving, etc. If these areas are subject to heavy rain, there may be potential for flash floods, mudslides, mud and debris flow, and other related hazards. One can list myriad other mountain/ wilderness rescue hazards for which local fire/rescue agencies should be prepared, including (but not limited to): landslides, rock fall, vehicles over-the-side, aircraft crashes, rail crashes, exposure to the elements, and simply getting lost.

Assessing urban hazards

Large modern cities are associated with almost limitless entrapment hazards. The character and nature of cities and the places where they are built help determine the rescue hazards of a particular area. In places where oil refineries and fuel storage are common, there is increased potential for confined-space rescue, high-angle rescue (e.g., on and within structures such as cracking towers), deep shaft rescue, explosions, and fires that trap workers on towers or other places where technical rescue may be required.

In Figure 1–4, a worker impaled through the thigh by a 1-in. diameter rebar winces in pain as he is triaged after being extracted from the collapse of a 3-million gallon reinforced concrete water tank under construction.

Fig. 1–4 Worker Rescued from a Concrete Water Tank

In places with heavy industry, there is increased potential for workers to be caught in machinery or crushed by machinery and products, such as rolled steel or pipes or high-piled stock. There is also more potential for confined-space rescue and for fires and explosions.

Where railways, airports, seaports, and major highways are located, there is increased potential for large transportation accidents requiring physical extrication operations. Where shipping and boating and recreation or commercial diving are common, there is increased potential for offshore rescue emergencies.

High-rise buildings present potential for window washers to become trapped in high places, for people to be trapped in elevators, or for high-rise fires that require firefighters to be placed onto rooftops to conduct SAR, evacuation, ventilation, and firefighting operations.

Urban areas bisected by rivers and flood-control channels are places where swift-water rescue emergencies may be common. Lakes and ponds in urban areas present a potential for dive rescue operations, ice rescue emergencies, and boating accidents.

Suburban areas may be the site of complicated extrication operations involving hands caught in lawn mowers and kitchen sink disposals, burglars trapped in chimneys, flash floods, or major auto accidents involving tractor-trailers and other heavy vehicles. Figure 1–5 shows firefighters searching for missing construction workers, 28 of whom fell nearly 80 ft when the roof of the tank collapsed beneath their feet as they were pouring fresh concrete. Several of the men were impaled on 1-in. diameter rebar. Mishaps of this nature are characteristic of the rescue challenges faced by firefighters in industrialized urban areas.

Fig. 1–5 Firefighters Search for Missing Construction Workers

One thing is sure: regardless of the terrain and other factors, denser populations offer more potential for people to become trapped. The local fire/rescue agencies have a responsibility to be prepared to manage the rescue emergencies that are likely to occur and to be ready for the unexpected.

Determining jurisdictional responsibility for rescue

An important aspect of rescue planning is to determine which agencies are legally responsible for rescue operations and to what extent their responsibility extends. It's also important to understand how different rescue agencies should intermesh with one another to ensure the most timely and efficient response. Understanding the statutes of individual states, federal laws, local charters, and orders by elected officials with respect to rescue service is critical. It's equally important to understand the exact language used in these mandates and to interpret the language in such a way as to meet the intent (as well as the letter) of local laws, charters, and statutes.

Some outdated state and county charters place primary responsibility for SAR with law enforcement. There are clear problems with this practice. First, few (if any) of these charters make a distinction between mountain/wilderness SAR and urban search and technical rescue. The charters often ignore the fact that fire departments have been conducting SAR operations in urban areas since cities first appeared in North America and that mountain SAR teams are usually ill-prepared to address the kinds of rescues that occur in the urban environment.

In addition, some county and state charters were written decades before the advent of fully paid (non-volunteer) fire departments and certainly before the appearance of modern fire department rescue capabilities. Many charters were written at a time when law enforcement agencies had the only fully paid public safety personnel who could be tasked with the responsibility to coordinate SAR operations (usually in addition to their law enforcement duties). In the days when many existing charters and statutes were devised, most fire department positions were volunteer and there was no formal rescue training for firefighters.

Consequently, in those early days, it was naturally assumed that law enforcement should be tasked with responsibility for SAR in perpetuity, and with no distinction between wilderness and urban environments or other pertinent factors. But the truth is, law enforcement rescue units respond to technical rescues in some city environments, even as the fire department is assigned to the same incident and with the same mission to conduct rescue operations. In many instances, there is a history of conflict about which agency (e.g. fire department or law enforcement) has responsibility and authority to manage the incident. One could accurately say that few existing county and state charters or statutes take into account the explosive growth of fire department-based rescue programs and urban SAR capabilities during the past two decades, and the result is a potential decrease or delay in service to the public, which rightly expects that the closest available rescue-ready resources will be dispatched when people are lost or trapped.

In short, many local and state regulations and laws ignore the modern reality that there is a clear distinction between mountain rescue and urban rescue and that many modern fire departments have developed highly effective rescue systems capable of managing both types of incidents. Some of these charters and protocols also ignore the fact that many fire departments—in their quest to ensure the readiness of firefighters to manage all technical rescues within their jurisdictional areas–have developed highly effective disaster SAR capabilities.

In Figure 1–6, who is responsible for rescue in this environment? Traditional assumptions about capabilities for wilderness SAR operations may not apply with the spread of fire department-based technical rescue capabilities. In many cases, it is the fire department that arrives first at rescue scenes like this one on a 10,000-ft mountain overlooking Los Angeles. The victim, who cannot be seen by the unaided eye, is trapped in an ice chute located in the middle of the photo, after having careened 500 ft off an icy hiking trail. This photo, taken by the author, who was riding as a rescue team leader in one of two fire department rescue helicopters searching for the victim, shows the complexity of some SAR operations and the extreme terrain in which some victims find themselves trapped. In an example of multi-agency cooperation under unified command, both the sheriff's department and the fire department responded simultaneously (as per the local county's protocols, which recognized both agencies as responsible for wilderness rescue). A law enforcement helicopter eventually located the victim, and the fire department helicopter from which this photo was taken performed a hoist rescue with assistance from other firefighters on the ground.

Fig. 1–6 Rescue Scene on a 10,000-ft Mountain

In many parts of the world, fire department units respond to both urban and wilderness rescue emergencies on a daily basis, and they conduct operations at a level equal to, or exceeding, volunteer mountain and wilderness SAR teams. In some cities, local law enforcement agencies staff technical rescue units capable of managing rescue operations in the urban environment, and they respond to these emergencies simultaneous to the fire department. Whereas conflict is often part of the history in these regions, there are options available that can allow the effective use of all available local resources in daily and disaster rescue emergencies.

In some of the most efficient systems, fire departments work closely with mountain and wilderness SAR teams and law enforcement to ensure a response that's as effective and *seamless* as possible. These teams operate at the scene under the umbrella of *unified command*; they may train together, conduct joint planning, establish rules to ensure that all appropriate agencies are notified immediately, and respond simultaneously to the same rescue incidents.

It is an important responsibility of local fire/rescue officials to clearly define the parameters of local response protocols and to provide clear guidelines for response-level personnel responsible for carrying out rescue operations. This is most important in places where there has been uncertainty over who is responsible and who is in charge of rescue operations.

Rescue Operational Planning

Long before he lost his life rescuing others in the collapse of the World Trade Center's twin towers on September 11, 2001, Deputy Chief Ray Downey, head of the FDNY's Special Operations Command, coined the term *rescue operational planning*. During his many years as a firefighter, lieutenant, and captain assigned to FDNY rescue companies, Downey developed and mastered the concept around which this type of planning is based.

Simply put, rescue operational planning is the process by which a fire/rescue agency (specifically, members of rescue or USAR companies) uses available intelligence, personal and institutional experience, training, and research and development, to make a comprehensive evaluation of local rescue hazards. The plan puts in place the resources and protocols necessary to address those hazards in a timely and effective fashion, implements effective rescue strategies based on that intelligence and planning, and ensures a reasonable level of safety for the rescuers. Rescue operational planning can be particularly effective during the most serious rescue emergencies, where *real-time* planning and implementation must often be conducted under highly dangerous and rapidly changing conditions.

Downey, who conducted and supervised more than his share of rescue operations in New York and across the United States during a stellar career that spanned four decades, divided rescue operational planning into the following twelve steps:

1. *Pre-incident planning and preparedness; knowledge of local hazards and local, regional, state, and federal response capabilities.* This central concept (discussed earlier in this chapter) emphasizes the importance of assessing, understanding, and quantifying the local rescue hazards and the total range of rescue capabilities available to the incident commander, including mutual aid, regional and state SAR teams, and FEMA USAR task forces.

2. *Development of advanced SAR capabilities.* This process refers to the need for fire/rescue agencies to establish the ability to manage rescue emer-

gencies that commonly occur in the jurisdictional area. One goal for a fire chief in this regard is to match his agency's rescue resources with the known local and/or regional rescue hazards. There are innumerable options and combinations and many existing models to accomplish this goal. The concept of developing advanced SAR capabilities is the subject of chapter 3.

Fire departments with special problems such as high-rise fires may need to consider unorthodox solutions such as specially-trained rescue or USAR companies that can be deployed onto the roof from helicopters to conduct search, rescue, ventilation, and firefighting operations.

In Figure 1–7, a USAR firefighter prepares to rappel from a copter onto the roof of a high-rise building.

Fig. 1–7 USAR Firefighter Preparing to Rappel onto a Roof

3. *Knowledge and use of internal and external resources.* Long before major emergencies threaten to overwhelm first responders, the best incident com-

manders have already made themselves familiar with the various special resources within their own agencies. These *internal* resources include rescue companies, USAR teams, hazmat units, heavy equipment units, etc. Good incident commanders also familiarize themselves with *external* resources, such as public works departments, law enforcement, building and safety, civilian contractors, the military, county and state offices of emergency services, the federal government, etc. These internal and external resources may prove useful in the event of a major rescue or disaster.

Without a basic understanding of the available internal and external resources (or without immediate access to technical experts who *do* know the information) and without a good idea of how best to use them, an incident commander operates with a blind spot that could sabotage his efforts to create a successful outcome during emergencies and disasters

4. *Size-up and survey of the rescue scene.* Effective size-up should be a basic skill of any fire or rescue officer. Assessing the emergency situation and understanding what one is seeing are two critical components of an effective incident action plan. The proper physical approach to an incident, the use of an *eight-sided view* of complex incidents, proper training and preparation to recognize the dynamics behind the emergency and predict the likely sequence of events if the incident is left unchecked, and the ability to focus on the main hazards while keeping track of the rest of the scene are important aspects of an effective size-up and survey.

5. *Assessment of the overall rescue problem and development of a corresponding rescue plan.* This step refers to the ability of the incident commander and other responders to understand the actual cause of the event, to recognize the set of problems that is now confronting them, and to develop

a course of action that will actually work to resolve these problems without unnecessary loss of life among the victims and without undue danger to the rescuers.

In Figure 1–8, a safety officer attached to a FEMA USAR incident support team observes as rescue squad members descend into the World Trade Center collapse. A problem with some urban search and technical rescue emergencies and disasters is the fact that the scene may be quite complex and dynamic, with unusual conditions of which the typical incident commander may have no previous knowledge or experience. This is where the issue of RPD^3 becomes critical. If incident commanders do not understand what they are seeing or do not correlate the scene before them with the dynamics that created the emergency, how can they be expected to choose the most appropriate course of action to save lives, protect personnel, and mitigate the emergency?

6. *Evaluation of safety considerations and development of solutions.* One of the primary responsibilities of firefighters, rescuers, and their officers, is to prevent unnecessary injuries or death to the responders. When firefighters and/or rescuers become lost, trapped, or seriously injured in the course of emergency operations, it has a deleterious effect on the responders themselves, who are now injured or dead, and their families. It also has a poisonous effect on victims who are in need of immediate assistance, but whose rescue may now be delayed because some responders must now divert their attention away from the original rescue problem to conduct a rapid intervention operation.

Addressing safety problems in a pro-active manner is critical to effective rescue operations. In Figure 1–9, a rescue squad attached to a FEMA USAR task force places shoring to prevent secondary collapse while SAR operations proceed inside the Pentagon after the 9-11 attacks.

Fig. 1–8 Assessing the Rescue Problem

Fig. 1–9 Addressing Safety Problems Proactively is Critical

It is far better to recognize the most serious life-threatening hazards from the outset and to deal with them in a rational way, than it is to have rescuers become lost, trapped, or injured later when the margin for error is razor-thin. Rescuers can work at their optimal levels when they feel assured that serious safety hazards are being addressed by the incident commander (IC) and other officers, when they trust their PPE, and when they know there is a plan to recognize and quickly address unseen or newly emerging safety hazards. Rescuer confidence, in turn, means the victims will be located and rescued sooner, with a higher degree of confidence in a positive outcome.

7. *Implementation of the rescue plan.* Embedded within the incident action plan of every major rescue emergency should be a rescue plan that lays out the most appropriate strategy and tactics for actually locating and extracting lost, trapped, or injured victims.

It is the responsibility of the incident commander, or, depending on the incident command system (ICS), the rescue branch or USAR branch director, the rescue group leader, the rescue team manager, the rescue squad leader, the rescue company captain or lieutenant, etc. to implement the rescue plan in a way that maximizes the available resources and ensures a reasonable level of safety for the rescuers. A good rescue plan will have several options in case the first strategy and tactics prove ineffective or conditions change rapidly.

Those charged with actually carrying out the tasks required to make the rescue plan work should be properly briefed, trained, and equipped for the job. They should be backed up by an effective rapid intervention plan that ensures immediate assistance in case things go wrong and rescuers become trapped, lost, or seriously injured.

The rescue plan must be constantly evaluated for effectiveness as it is carried out and appropriate adjustments made to ensure the main objectives are met.

8. *Coordination of resources and operations and coordination among team members.* The job of incident commanders, or the person they designate to supervise rescue operations (whom we'll call the *rescue commander*), is not unlike that of the conductor of a symphony orchestra. Typically, rescue commanders have at their disposal a number of resources (instruments) whose actions must be coordinated, controlled, and choreographed to prevent conflicting actions and to bring about a harmonious resolution (music). Whereas the conductor must understand the music and must direct the timing and tempo of the orchestra and quickly react to unexpected glitches (such as out-of-tune instruments or broken strings) to achieve the aims of the musical piece, rescue commanders must exhibit similar mental agility. They must understand the conditions that caused the rescue situation, they must direct the timing and tempo of the rescue response, and they must quickly react to unanticipated events. Events such as equipment malfunction, secondary events (e.g., secondary collapse, explosion, etc.), and a wide range of other tasks must be conducted simultaneously in order to successfully rescue victims.

If rescue commanders are unable to multi-task or are not prepared to coordinate multiple (sometimes disparate) resources in concert with the mission of locating, freeing, and removing victims, they will find themselves in trouble. The rescuers and victims that the commander is supposed to help may find themselves in even worse trouble. This is where experience and a good command presence are key to accomplishing the goal of rescue.

Experienced rescue commanders, like experienced fireground commanders (who may be one and the same) know how to get things done under duress, how to avoid mistakes that can lead to disaster, how to use the knowledge and skills of technical specialists to the best advantage, how to ensure redundant safety for the rescuers, and how to find little shortcuts that avoid pitfalls and delays. Less experienced rescue commanders, who may not have been afforded the opportunity to gain years of experience commanding and supervising major rescue operations, may supplement their experience and skills through the use of effective and realistic training, case studies, and simulations.

9. *Effective communications.* It is self-evident that effective communications are critical to the outcome of a rescue emergency. In this sense, communication begins with the use of an ICS that allows the rescue commander to establish clear and unambiguous command and control. This control continues through the use of effective verbal and visual communications, the availability and use of good radio communications, and even the use of ancillary communications. Ancillary communications might include remote monitoring, aerial command with ground links, visual aids such as thermal imaging and night vision that allow the commander to observe real-time operations under adverse conditions, and other forms of communication.

Effective rescue operations rely on clear identification of personnel and positions and a mix of verbal, written, hand, and radio communications to get the job done under fluid circumstances (Fig. 1–10).

Fig. 1–10 Effective Communications Get the Job Done

Inherent in effective communication is the ability of individual rescuers to maintain unimpeded communications with their leaders, and through them, the rescue commander. Rescuers must keep their commanders apprised of conditions, progress, and needs. They must highlight critical safety and operational concerns to ensure an immediate rapid intervention operation if things go wrong. Conversely, the ability to maintain communications up and down the chain of command will augment the rescue commander's ability to determine whether or not the strategy chosen is being carried out through the use of effective tactics by rescuers and commanders at every level.

As the 9-11 terrorist attacks so dramatically illustrated, communications is essential to ensure that personnel receive critical safety and operational information throughout the course of the emergency. This becomes especially important when critical events occur, or when the rescue commander decides that it's necessary to conduct an operational retreat because conditions are rapidly deteriorating. The aftermath of the 9-11 attacks also illustrated the criticality of communications when personnel become lost, trapped, or seriously injured at the emergency scene. Rapid intervention operations are highly dependent on good information about where personnel are trapped or lost, the conditions they are facing, the tools and manpower needed to locate and extract them, the best access to reach them, and any continuing hazards.

10. *Effective command and control.* Ray Downey understood that effective command and control is one of the central themes of rescue. He understood that, without good coordination and direction by the incident commander (and his designated rescue commander), a situation that's already fraught with danger can be made much worse. This is why he made it a tenet of rescue to establish a command-and-control system that works.

In the modern world, the use of a validated incident management system is critical to the ability of rescuers to handle complex, large-scale, lengthy, and high-risk rescue emergencies and disasters. Without the use of a viable ICS, the rescue commander may find himself making decisions without the necessary information. He may find that his strategic decisions are not translated into effective tactics to achieve them, and he may find that there is little or no personnel accountability. Further, he may find himself (and his personnel) at the mercy of changing conditions without the command and control to address them. Simply put, good command and control is essential for positive outcomes to major rescue emergencies.

11. *Innovative capabilities—the ability of fire/rescue personnel to improvise innovative solutions for complex rescue problems.* The ability of firefighters and rescuers to innovate and to adapt quickly to changing conditions was a central theme of Downey's approach to rescue operations. Innovation and imagination were hallmarks of the Ray Downey methodology, and they served well the scores of people rescued through his efforts and those of the rescuers he commanded and coordinated. Extrapolating to the plethora of rescue

conundrums that will confront firefighters and rescuers in the coming years, it's clear that innovation, improvisation, and imagination will be essential to successful outcomes and more effective approaches to new rescue problems.

12. *An effective contingency plan that anticipates the unexpected and ensures that there is a plan B, plan C, and plan D ready for immediate implementation if plan A fails.* The most experienced rescuers and rescue commanders understand that things often go wrong or that unexpected events often complicate the situation during the course of complex rescue operations, even when the best-laid plans and strategies are implemented. Rescue commanders who fail to back up their primary strategy with a number of alternative plans ever ready to implement if or when complications arise or things go astray, are leaving themselves and their personnel, as well as victims awaiting their assistance, vulnerable to the vagaries of chance.

Based on years of experience with complex rescue emergencies and disasters, it has become this author's practice when asked to estimate the time required to complete the rescue operations, to *guesstimate* that time and then double it. The reason is simple: in case after case, unexpected events, unforeseen complications, equipment malfunctions or delays, and the occasional human error, have caused rescuers to resort to alternate plans and have essentially doubled the time originally estimated for the extraction of trapped victims.

When operating at the scene of urban search and technical rescue emergencies, it has become this author's practice (and the practice of his colleagues) to always have a plan B, plan C, and sometimes plans D and E in place or being prepared. This is a standard tenet of many rescue teams and fire departments. It's an approach that serves well the victims waiting our assistance.

Conclusion

Planning for daily rescue emergencies is essential for effective management of the predicaments in which people tend to find themselves on a regular basis. It is part of the risk-management responsibility of all local fire departments and rescue agencies. Proper planning for daily rescue operations has an added benefit: it lends itself to planning for disaster rescue operations, as well as development of multi-tiered rescue response systems that are prepared to handle both daily and disaster rescue.

Endnotes

[1] If significant rescue hazards are identified, it is an indication of the possible need to develop a formal rescue program, including rescue training for personnel, procurement of PPE, purchase of technical rescue equipment, development of response plans, and (if necessary) implementation of mutual aid agreements.

[2] Arizona crossing is a term commonly used in the west to describe the location where a road, driveway, or path crosses a shallow or dry river bottom. The term is generally used in reference to desert regions, but may include any place where people routinely cross dry or shallow rivers or streams. Arizona crossings are frequently the site of fatal encounters when people tempt fate by traversing them during storms, when the current is faster, deeper, and more likely to displace the weight of automobiles, causing them to be swept away. Because they are prone to flash flooding, Arizona crossings are a key indicator of flood and swift-water rescue hazards, and they are often the site of high-risk rescue attempts. Note: some Arizonans may prefer the term California crossings.

[3] Recognition-primed decisions refer to metaphorical "slide shows" (embedded in the mind of the incident commander through actual experience or training simulations) that correlate various emergency scenes with possible outcomes. See chapter 5 (Training) for details.

2

Planning for Rescue-Related Disasters

Disasters, by nature, are often capricious events that can leave trails of destruction even in places that aren't accustomed to large-scale emergencies. A hallmark of some disasters is a "hopscotching" effect, whereby entire neighborhoods may be wiped out, yet one or two homes or other buildings in the middle remain untouched. Other disasters may result in complete wipe-outs of neighborhoods, towns, or even cities. Even then, there are often pockets of survivors in the most unlikely places. While some disasters are memorable for the capriciousness, most are not totally unpredictable. To the contrary, many disasters can be anticipated and effectively managed if appropriate plans are in place to deal with likely and potential hazards. A key to successful rescue operations in a disaster setting is effective disaster planning.

We have already established that rescue is one of the most basic missions of the fire service, and that fire departments and rescue agencies have a responsibility to develop and implement rational plans to manage daily rescue emergencies in a timely and professional manner (thereby providing the best chance of survival for victims who are trapped or lost). We've established that multi-tiered rescue systems consisting of well-trained first responders supported by rescue companies, USAR task forces, fire/rescue helicopters, and other specialized units, are an effective way to manage everyday urban search and non-urban technical rescue emergencies (see chapter 3 for guidelines on establishing multi-tiered rescue systems). Now it's time to examine the planning process for rescue-related disasters; how hazard identification is critical to disaster response; how we can plan to use systems originally established to manage "daily" technical rescues to handle the largest disasters; and how the planning process helps ensure the more effective use of specialized disaster rescue during disaster-sized rescue emergencies.

In Figure 2–1, a mud and debris flow that swept down an Italian mountain demonstrates the capriciousness of disasters. Were local fire/rescue officials aware of the potential for such an event? Had they developed response plans for a major mud and debris flow to strike their town? Were the local firefighters and rescue teams trained, equipped, and prepared to assess the situation, request additional resources, and begin the process of SAR? Was there a smooth implementation of mutual aid response? Can it happen again, in another city, or another nation? Can it happen where you work?

Fig. 2–1 Mud and Debris Flow in Italy (Photo by Plinio Lepri, Associated Press)

In addition to their often capricious nature, disasters also have a tendency to compress many serious fire/rescue/EMS and hazmat emergencies in a narrow timeframe that can overwhelm the capability of local, regional, state, or even federal response systems. This often forces firefighters and other rescuers to confront a multitude of urgent situations simultaneously, which in turn causes competition for scarce resources under the most demanding conditions, resulting in worst-case scenarios for victims and emergency responders alike.

Figure 2–2 shows L.A. County Fire Department (LACoFD) and city firefighters breaching and tunneling their way through a collapse to reach a man trapped in his truck, crushed on the first floor of this three-story parking structure during the Northridge

earthquake in 1994. It required nine hours to extract the man from his entombment. Does your agency protect earthquake-prone areas or other disaster hazards of similar scale? If so, is there a plan to conduct effective and timely SAR operations under conditions like this?

Fig. 2–2 Firefighters in Northridge Earthquake (Photo by Mike Meadows)

For example, firefighters and rescuers in rural areas, who might physically extract trapped victims only a handful of times during the course of a career, and even those assigned to busy urban areas (who might conduct dozens of physical rescues every year or two) can find themselves faced with hundreds or even thousands of people requiring physical rescue, all at the same time, and all within a compact geographic area. Such is the nature of disaster: the equivalent of a lifetime of rescue experience can sometimes be compressed into a period of hours, days, or weeks.[1]

In chapter 1, we discussed the need for multi-agency cooperation and the use of specialized resources like lifeguards, the U.S. Coast Guard, mountain SAR teams, heavy equipment, canine search teams, and other mission-specific resources during daily rescue operations. And we've established that fire departments have been formally designated the agency primarily responsible for conducting urban search and technical rescue (as well as EMS, hazmat, and firefighting) operations in many parts of the world. When daily rescue emergencies become disasters—or when they begin as disasters from the outset—many of the same principles of rescue apply, only on a much larger scale.

The essence of disasters (after all, disasters by definition are so large or complex that the capabilities of the local jurisdictional agencies are overwhelmed) necessitate effective disaster response plans that address rescue-related consequences in a timely and effective manner. According to this line of thinking, local fire/rescue agencies have a responsibility to plan for effective response to disasters that leave many victims trapped, lost, or otherwise in need of physical rescue.

Step 1: Establishing Agency Responsibilities

In chapter 1, we discussed the need to determine legal and procedural responsibility with respect to which agency (or agencies) manage rescue operations. Likewise, the first step in developing a disaster plan is to establish which agency has primary responsibility for SAR operations in the disaster setting, as well as clarifying which agencies play a support role. This is an

equation that can differ according to the statutes of individual states, federal laws, local charters, by orders of elected officials, or by agreements of a group designated to address such issues.

Who is responsible for SAR when disaster strikes your community? (Fig. 2–3) The disaster rescue planning process is a good time to consider the need to update charters and statutes to reflect the modern reality *on the ground*. It's critical for all involved parties to understand from the outset which agency will take the lead in disaster rescue operations and which agencies will support the lead agency when a particular type of disaster strikes. Without this understanding, and without charters and statutes that ensure commensurate (and legally mandated) authority, responsibility, and funding, the disaster planning process may be flawed from the beginning, and the result may be a double disaster (e.g. the actual disaster occurs, and then the lack of planning causes a disastrously poor emergency response).

Fig. 2–3 Delegating Responsibility

During the course of this writing, we saw examples of this dynamic play out during southern California's October 2003 firestorms (to which this author and thousands of other firefighters responded for days at a time), which killed more than 20 people, caused the loss of more than 3,500 homes, and burned more than 750,000 acres. In the aftermath of the worst wildfire disaster in California history, it became evident that the responsibilities of certain agencies were not well defined, that the approval process for special equipment and tactics had large holes, that many fire departments in San Diego County were grossly underfunded and understaffed, and that recommendations from lessons learned in previous fire disasters had been ignored by local and regional civilian policy makers. As a result, there was speculation among the public, elected officials, and the news media about the fires. Some believed that some fires might have been contained before spreading into neighborhoods. Instead, they grew too large for the short-staffed firefighting forces (who were then being spread even more thin because simultaneous wildfire disasters elsewhere in the state required resources to be dispatched to assist other counties). Ultimately the results included the loss of many hundreds of structures that otherwise might have been saved.

In terms of rescue-related disasters, we saw similar situations play out when the Loma Prieta earthquake struck the San Francisco area in 1989, and when Hurricane Hugo struck the East Coast the same year. In both events, elected officials, the public, and the news media noted that the Federal response was slow, unwieldy, and not well-prepared to manage large-scale SAR operations. The formation of the FEMA National Urban Search and Rescue Response System (including the development of 28 FEMA USAR Task Forces based in major fire departments across the United States) was a direct result of these two events and the intense criticism about the lack of disaster rescue planning and response.

In one sense, the problems encountered during the Loma Prieta quake and Hurricane Hugo resulted in a huge improvement in the government's readiness for major rescue-related disasters. Ultimately, the resulting systematic improvements allowed the U.S. government to respond faster and more effectively to events like Hurricane Iniki (1992), the Northridge Earthquake (1994), the Oklahoma City bombing (1995), the 9-11 attacks on the World Trade Center towers and the Pentagon, and a number of other rescue-related disasters. But it's not always necessary to experience a disaster yourself to employ effective disaster planning methods and to be ready for the next rescue-related disaster to strike your region. We can get ahead of the power curve through effective disaster rescue planning, and that is the goal of this chapter.

In recent years, the fire and rescue services have vastly improved their capabilities for the management of rescue-related disasters. Since the 1980s, there has been a virtual explosion in the number and scope of formally organized urban search and technical rescue teams and units around the world. For example, in the United States this includes rescue companies, USAR task forces, SAR teams, and other rescue-related units established by local professional and volunteer fire departments. It includes expansion of the scope and capabilities of government-sponsored and civilian rescue teams and systematic organization and coordination of these teams at the county, regional, state, and federal levels. Improvements also include the previously mentioned development of 28 FEMA USAR task forces, as well as two internationally deployable USAR task forces under the auspices of the U.S. Agency for International Development's (USAID) Office of Foreign Disaster Assistance (OFDA).

Effective disaster rescue plans should include provisions to bring all these forces to bear when a rescue emergency is of such size and scope that it overwhelms the capabilities of local resources (the very definition of a *disaster*). In basic terms, when the resources of a city or town are overwhelmed by a disastrous rescue emergency in the United States, it often becomes the responsibility of the local IC to request mutual aid from adjacent communities (or the jurisdictional county or region), which in turn will send additional resources, including specialized teams, where they are available.

At the moment it becomes evident that county or regional resources are overmatched by a disaster, the state may be requested to step in and dispatch additional (statewide) mutual aid resources and specialized teams (where available). Finally, if the situation is too large or complex for state resources to completely control, the jurisdictional state (governor) may request federal assistance, which is the trigger for implementation of the Federal Response Plan and the appropriate federal emergency support function (ESF). The ESF most closely related to rescue disasters is known as *ESF-9*, the *Urban Search and Rescue Emergency Support Function*.

FEMA is designated as the lead U.S. government agency when ESF-9 is activated for disasters in the United States, and more than 26 other federal agencies are charged with responsibility for support roles under the direction and request of FEMA. A key resource within ESF-9 is the previously mentioned system of 28 FEMA USAR task forces strategically located across the United States.

If there is no obvious and visible fire department or rescue team presence, citizens will take matters into their own hands (Fig. 2–4). When local fire and rescue agencies are overwhelmed, assistance from the public may be required. Therefore, effective disaster plans include protocols that allow fire and rescue personnel to organize citizens into effective convergent responder teams in disasters that overwhelm local rescue capabilities.

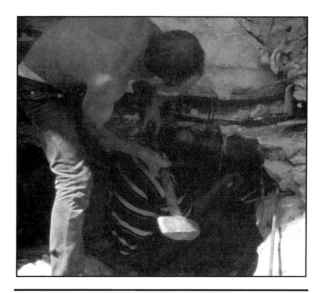

Fig. 2–4 SAR Operations in the 1999 Turkey Earthquake (Courtesy of AKUT, Turkey)

It's important to emphasize that U.S. state and federal rescue resources will generally report to the local IC, often the local fire chief of the city, town, or county affected by the disaster, and these resources will support the needs of the local IC. The point is that in the vast majority of disasters, local authorities retain command and control of the disaster, with the state/federal government providing whatever assistance is needed and requested.

To effectively utilize these resources, it is incumbent on the fire chief and other local authorities to be thoroughly familiar with regional and state response and mutual aid systems and with the Federal Response Plan. If, for example, the local fire chief and his commanders (down to the company level) do not understand the composition, capabilities, limitations, and basic operational parameters of the FEMA USAR task forces, they may find themselves ill prepared to request and utilize these highly advanced, highly experienced, disaster-hardened teams of firefighters and rescuers. This is an issue that demands attention at the local, regional, and state level, so that the most effective assistance can be applied to people in need of SAR in the midst of future disasters.

In the event of terrorism-related disasters, the normal system of mutual aid may be preempted to address national security/homeland defense issues. For example, the disaster site automatically becomes a crime scene (at the same time that it is a rescue scene); and the Office of Homeland Defense (including the FBI and other federal agencies) will automatically have jurisdiction of overall *crisis management* and *consequence management*, in conjunction with the responsibilities and roles of local government.

In basic terms, regardless of whether or not local officials request outside assistance at a potential terrorism attack that has become a disaster, assistance will be coming in the form of homeland security resources and authorities. With the threat of terrorism looming in the coming years, it is important for rescue commanders to understand this aspect of jurisdictional responsibility.

Step 2: Hazard Assessment

Once the issues of responsibility and authority for disaster planning and response have been addressed, the next step in disaster planning is to conduct a hazards assessment, or an inventory of the disasters likely to strike a given area. Disaster potential can be evaluated on the basis of local disaster history; on an assessment of the local terrain;[2] local weather patterns, including hurricanes, tornadoes, and other weather-related factors; man-made features,[3] including transportation hubs such as airports, rail yards, and shipping ports; current political and social factors (which are closely related to domestic and international terrorism); as well as future changes in the local landscape, man-made features, social/political forces, and other factors related to the potential for future disasters.

A realistic assessment of local risk enables fire/rescue agencies to plan effective responses to events that may plausibly occur in a given region or city. *Realistic* is the key word. Some disaster plans have been woefully inadequate because the risks were hugely underestimated out of ignorance of local hazards, a lack of scientific and historic data from which to make calculations, the desire to put the best face on a known hazard to protect business, tourism and other local interests, and simple apathy.

Other disasters, including terrorist acts or the failure of engineered structures, simply have not been anticipated at all. Not every disaster can be predicted with a guarantee of accuracy. However, there are ways to deduce the potential level of disaster from studies of historical records, geological data, weather patterns, and a rational examination of man-made hazards.

Using historic data to facilitate hazard assessment

Local history is generally—but not always—a reasonable indicator of the potential for future disasters. Where earthquakes, tornadoes, hurricanes, floods, and other disasters have happened in the past, there's a general assumption that they can happen again in the same places as long as geography, weather, and other conditions remain somewhat consistent. Times change of course, and new and unseen forces such as climactic changes, industrialization, and even terrorism occasionally change the patterns of disaster in a given region.

Every disaster is a storehouse of lessons for those in search of them. The problem is determining how society should weave these lessons into the fabric of modern daily life, commerce, and governance in a manner that will prevent the unnecessary loss of lives and property when the next disaster strikes.

This problem was illustrated in the wake of the 1994 Northridge earthquake, which devastated parts of southern California and was (at the time) the most expensive disaster in U.S. history. What lessons were really learned from Northridge, and how have they been applied? To some, the main significance of Northridge was its colossal cost. To others, it was vindication that California's building codes and structural retrofitting requirements for older buildings—among the most stringent such standards in the world—protected the populace from shaking that decades earlier would have killed hundreds or perhaps even thousands of people in Los Angeles. To still others who were more personally affected by death to family and friends, the Northridge quake was their nadir; they could not have imagined worse loss.

But to experienced firefighters, rescuers, and commanders (who understand that a bigger disaster and a different threat may lie just around the corner or with the next rising of the sun), Northridge was a sort of *warning shot.* They view the event as a reminder to expedite preparations for far more devastating earthquakes that loom on the horizon, not just in L.A., but also in far-flung places like Anchorage, San Francisco, San Diego, British Columbia, and Oregon, along the New Madrid fault zone, and across other seismically active regions of the world. Even in New York City, where earthquakes are infrequent, significant faults have been discovered, and one powerful jolt could cause immense disaster, exceeding even that of the 9-11 attacks.

As if to punctuate the lessons of the Northridge quake, thousands of people were killed exactly one year later by a quake that rocked the port city of Kobe, Japan, proving once again that modern, Western-style construction (even steel-frame buildings) might fail when the ground shakes long enough and with sufficient violence.

With all evidence to the contrary right under their feet, it's interesting to note how many Westerners seem to observe disasters in other nations with a safe sense of detachment, apparently lulled by constant assurances that our methods of erecting buildings and freeway overpasses are somehow superior to those that fail with regularity in other modern industrial countries. Americans may believe that the post-quake conflagrations that swept Kobe and Tokyo are an Asian phenomenon rooted in narrow streets and wooden homes built close together and that efforts to retrofit non-reinforced masonry buildings are somehow a guarantee against catastrophic failure. We may think that the parking structure at the local mall won't pancake down, layer upon flattened layer or that standing in the doorway will protect us if our apartments are crushed by upper floors when the *soft* first story collapses during a quake or that the local fire station will withstand the shaking and there will be plenty of fire engines and rescue teams to snuff the blazes and locate and remove those trapped in the rubble.

Many people in the West—including some fire/rescue professionals who should know better—have been left with a vague sense that scenes of quake-spawned devastation, or the consequences of terrorist events they've watched on the news, are strictly a foreign phenomenon, and that it can't happen here. In some ways, they may be the fortunate ones because who among us wishes to live every waking moment in fear that the walls around us may come crashing down at any moment?

The truth is, no engineer or building contractor in the world can guarantee that a particular apartment complex, school, freeway overpass, shopping mall, parking structure, office building, or high-rise will survive the next big earthquake. And, as we tragically discovered when the 9-11 attacks occurred, there is no guarantee that existing buildings will withstand the consequences of major terrorist attacks.

Hurricanes as a barometer for hazard assessment

Hurricanes provide a good working example of the processes required to accurately evaluate local disaster hazards. Hurricanes provide a good case study for the necessity of heeding historic data when planning for disasters. The hazard is clear in places like the United States, where more than 60% of the population lives in coastal states including Maine, Texas, Hawaii, and Puerto Rico, and many of these regions are subject to hurricanes.

Tidal surge is often misidentified as the most danger-ous killer related to hurricanes, followed closely by strong winds, flying objects, and the kind of structural damage that often characterizes hurricanes. But there's a missing factor that most people, including many of those responsible for disaster planning and response, don't seem to acknowledge: over the past 30 years, neither tidal surge nor structural collapse caused the most fatali-ties in hurricanes. The true culprit in terms of deaths has consistently been inland and coastal flooding caused by the intense rain that accompanies hurricanes.

Unlike flood surges, inland flooding can occur hun-dreds of miles from the coast. Sometimes, it's not the largest or strongest hurricanes that cause the worst damage and loss of life. In fact, it is often smaller or weaker hurricanes that cause the worst casualty rates, especially those that drift slowly or stall over land, particularly land that's covered in hills and mountains, where runoff may be most deadly.

Examples of the danger of inland flooding during hurricane events abound, including the following sam-pling from only three decades. In 1955, hurricane Diane caused more than 200 deaths, most of them from inland flooding. In 1972, hurricane Agnes took 122 lives, most from flooding that occurred long after the hurricane died out. Tropical storm Claudette (1979) dumped more than 45 in. of rain in parts of Texas. In 1994, 33 people drowned when inland flooding struck parts of Georgia in the aftermath of tropical storm Alberto. During hurricane Floyd (1999), 50 of the 56 fatalities were linked to inland flooding. And then there was tropical storm Allison, which surprised experts by killing more than 50 people in Texas and other affected states without ever reaching hurricane strength.

It has taken nearly three centuries of hurricane disasters for the majority of decision-makers to finally acknowledge that the worst killer in hurricane disasters is no longer tidal surge or high winds, but the immense downpours that accompany hurricanes

and tropical storms. Finally, there is strong evidence that government is moving toward the concept of improved readiness as the key to managing the worst life threats from hurricanes and tropical storms, even as it continues addressing other serious hurricane-related problems like structural collapse, injuries from wind-driven debris, and tidal surge.

According to many researchers, hurricanes are becoming a worse threat to human life with each passing year—at least in the near future. In July 2001, scientists announced that North Atlantic hurricane activity doubled since 1994, and that this high inci-dence of hurricanes may last until around 2041. In a research article published in the journal *Science,* researcher Stan Goldberg of the U.S. National Oceanic and Atmospheric Administration (NOAA) was quoted as saying, "We believe a worst-case disaster is waiting to happen in the United States. We're talking about repeated multi-billion-dollar damage and perhaps hundreds or even thousands of deaths."

Goldberg cited studies showing a 40-year cycle of natural increase in the temperature of the sea surface, which in turn leads to fiercer storms that eventually become deadly hurricanes. From 1971 to 1994, the North Atlantic averaged 1.5 major hurricanes per year.[4] Since 1994, there has been an average of three major hurricanes per year in the North Atlantic. This is a problem (in part) because the previous 40-year cold period along the Atlantic and Gulf coasts marked a time of unrestrained coastal development in hurricane-vulnerable areas.

The Sea of Marmara rushed inland like a tsunami and swamped coastal towns when the land subsided during a devastating earthquake that struck Turkey in 1999 (Fig. 2–5). Do your disaster plans take into account flooding on this scale that could result from heavy rains, tsunamis, land subsidence, levee failure, dam collapse, or hurricane-related tidal surge?

Fig. 2–5 Devastating Earthquake in Turkey (Courtesy of AKUT, Turkey)

equipment (including PPE), effective disaster plans that include the use of mutual aid, and protocols for requesting local, regional, or state resources. These resources may include swift-water rescue teams, helicopter-based rescue teams, National Guard units, urban SAR teams, dive teams, or FEMA USAR task forces. In some cases it may be prudent to create new resources such as rescue companies, local USAR task forces, swift-water rescue teams, and helicopter rescue units, or it may be necessary to augment the staffing of existing rescue resources.

Earthquakes and hurricanes are just two examples of disasters that can be anticipated through effective hazard assessment. It's incumbent on local fire/rescue authorities to determine the particular disaster hazards that affect their jurisdictions and to conduct realistic assessments of the potential consequences for which an effective response must be planned.

Step 3: Assessment of Available Rescue Resources

Once an accurate assessment has been conducted and documented, it's time to evaluate the local fire/ rescue/EMS and hazardous materials response capabilities and their readiness to manage the consequences of possible disasters identified in the hazards assessment (see chapter 3, Developing Multi-Tiered Rescue Systems, for details on specific types of rescue and USAR resources and systems).

If the resources required to handle anticipated disasters are found to be lacking, the next step is to build up adequate strength by providing training, appropriate

Step 4: Writing the Disaster Rescue Operations Plan (DROP)

DROPs should be written, distributed to all concerned parties, discussed among all participating agencies and all participants in the system, and changed according to new lessons or changing conditions. DROPs should be developed as *living* documents, subject to periodic review and revision to ensure that they address the most likely scenarios in a realistic way.

In the United States, DROPs should be consistent with the ICS, created by the Firefighting Resources of California Organized for Potential Emergencies (FIRESCOPE) in the 1970s to address exactly the types of problems that are endemic to disasters resulting from wildfires, floods, earthquakes, hurricanes, tornadoes, explosions, transportation accidents, and practically any other disaster scenario.[5] DROPs should also conform with the SEMS, which incorporates certain incident command and emergency management principles into the operations of all government agencies involved with the response to a disaster or other major incident.

For example, in places like California, the use of SEMS is mandated by law.[6] The California SEMS law mandates that all government agencies cooperate with each other and use the same terminology in the event of any large emergency or disaster that causes the jurisdictional agency to request outside help. SEMS also includes the development of multi-hazard functional plans (MHFP) that clearly spell out the responsibilities of all responding agencies and government entities (including public works, fire departments, law enforcement, lifeguard agencies, as well as city, county, and state governments) for any type of disaster that's deemed possible in a given jurisdiction.

As of this writing, the U.S. Department of Homeland Security is considering adoption of rules mandating the use of SEMS by public safety agencies nationwide. Even where SEMS isn't yet legally mandated, it is appropriate to incorporate SEMS protocols into disaster plans. The advantage of adopting SEMS and ICS is that it sets the stage for an organized response of all local, regional, state, and federal resources and ultimately improves their ability to protect life, the environment, and property with the highest degrees of personnel safety and accountability.

The MHFP, which can be adopted at the city, county, regional, or state level, provides a guide for managing specific types of disasters. It identifies the lead agency for each type of disaster anticipated and lists all the supporting agencies that will report to the lead agency. Typically, the fire department is listed as the lead agency for any disaster at which urban search and technical rescue operations, fireground operations, EMS, and hazmat response will be a primary concern. Depending on the jurisdiction and its specific hazards, the MHFP may cover disasters such as earthquakes, floods, dam failure, tsunamis, hurricanes, tornadoes, terrorist attacks, explosions, structural collapse, commercial airliner crashes, train derailments, shipping accidents, riots, and practically any other disaster that might occur in a given place.

It's also necessary to understand the federal response plan, which identifies lead agencies and supporting agencies for different types of disasters. The federal response plan considers different types of disasters, with corresponding ESFs that place certain federal agencies in charge of the federal response to any disaster at which federal assets are requested or dispatched across the United States and its territories. More information about the federal response plan may be obtained at the website of the FEMA at www.fema.gov.

Without the use of an ICS, SEMS, or MHFP, and without an understanding (and readiness to use) the federal response plan, the local DROP will be flawed from the beginning, and a stream of avoidable problems may follow. For victims trapped in collapsed buildings, stranded by floods, or trapped by any other type of disaster, the presence of a viable and well-tested disaster plan that ensures timely and effective rescue response may make the difference between life and death.

Step 5: Testing the DROP

Step 5 in the disaster planning process is to test the DROP using local and regional resources under simulated disaster conditions. This can be accomplished by disaster simulations, multi-agency drills, or even didactic (e.g. tabletop) sessions where the plans are brought under review by highly experienced rescue personnel and others with decision-making responsibility. Another option is to conduct case studies of actual disasters, with an emphasis on determining what worked and what didn't, and to evaluate the effectiveness of like resources and systems under actual disaster conditions.

Step 6: Adjusting the DROP

If the testing step indicates that the local systems are inadequate to effectively manage likely disaster scenarios, the natural next step is to make appropriate adjustments. Adjustments may include changing the responsibilities of local agencies (including identifying a new primary responsible agency), reassessing local disaster hazards, providing additional training, procuring better equipment, creating new rescue resources, augmenting the staffing of existing rescue units, improving mutual aid plans, and establishing or improving protocols for requesting state and federal resources during times of disaster. After improvements have been made, it's time to test them again and make appropriate adjustments.

Common Disaster Rescue Hazards

Disasters, like daily rescue emergencies, can be predicted (if not always prevented). If one or more of the following hazards can be found within (or near) a fire/rescue agency's jurisdiction, the corresponding potential for USAR-related disasters should be addressed by plans to effectively manage them.

- Natural rivers or streams
- Steep mountains
- Large industrial complexes, including explosives and military installations
- Sea ports, harbors, and piers
- Large amusement parks
- Nuclear power plants
- Potential terrorism targets such as embassies, churches, mosques, and government buildings
- Landslide areas
- Earthquake faults
- Flood-control channels (indicating serious flood hazards)
- Refineries
- Water treatment plants
- Airports
- Mines (operating and non-operating)
- Tunnels used to transport numerous people through mountains and beneath cities
- The ocean
- Unstable coastal cliffs and bluffs, subject to collapse
- Tornado conditions
- Above-grade levees (indicating serious flood threat)
- Dams
- Coastal subduction zones that may generate tsunamis and/or cause the land to subside
- Offshore thrust faults that may generate tsunamis
- Large construction projects
- Railroad tracks, switching yards, and other rail facilities
- Subways
- Regions subject to flash floods, rockslides, or mud and debris flows
- Coastal areas (especially those at or below sea level) subject to hurricanes
- Coastal zones subject to undersea landslides in deep offshore canyons that can cause tsunamis
- Cities in proximity to active volcanoes and the drainages that carry volcano-spawned mud and debris flows

Emerging Disaster Hazards

Disaster planning, by its very nature, is a dynamic process that never ends. As society evolves, as weather patterns change, as demographics and politics shift, and as the terrain changes, the conditions that lend themselves to disaster also change. Terrorism is a clear example of the dynamism of disaster planning and response. For example, as recently as two decades ago, terrorism was—in the minds of most Americans—something that happens in other places. Even ignoring the persistent threat of domestic terrorism that has existed to some degree since long before Abraham Lincoln was assassinated more than a century ago, direct attacks on American soil by terrorists from foreign lands were considered so unlikely that international terrorism was not even on the radar screens of most fire/rescue agencies.

The 1993 bombing of the World Trade Center changed all that for many cities and states. It was one of the clearest signs of the emerging threat of international terrorism on U.S. soil. It should have served as a wake-up call for every major fire/rescue agency, but many chose to ignore the evidence.

Case study 1: California mud and debris flow disasters

Assessing the potential for disaster to strike a particular city, county, region, or state, can be problematic because natural and man-made disasters are hard to predict and because people by nature tend to downplay the possible consequences of disasters, even when all evidence points to the contrary. Many people simply don't want to believe (or acknowledge publicly) the possibility that their community may be struck down by the full force of an earthquake, hurricane, tornado, landslide, avalanche, dam failure, or industrial disaster, and fewer still are ready to consider the true potential for foreign and domestic terrorism.

Ask people whose towns are literally built on the edges of giant earthquake faults that have already caused disaster in the not-so-distant past, or whose towns have already been razed and rebuilt after some calamity that may strike yet again, and many will tell you flat out that the odds of recurrence are so remote that they just don't think about it. Some will say they put their fate in God's hands. Those with a more fatalistic view may reply with a shrug, "If it happens it happens." Still others simply refuse to admit they are living on the precipice of calamity, even when signs portending disaster lay all around. Such is human nature.

An example of this human tendency is the struggle to prepare for devastating mud and debris flows in California, the most disaster-prone state in the United States. During the 1990s, California averaged more than 33 federally declared disasters each year, including earthquakes, floods, wildfires, urban interface conflagrations, snow storms, and riots. It's a place where disasters are part of the background noise of daily life. The rugged terrain, wildfires, floods, unstable geology, volatile effects of plate tectonics, cycle of deadly storms, demographics, and the potential for terrorism and civil unrest practically guarantee a continuing cycle of disasters for the foreseeable future.

After managing the consequences of deadly floods from 1992 to 1997; after living and working through the 1992 riots that burned thousands of buildings across L.A. County; after surviving the firestorms of 1993 that burned hundreds of homes, killed several people, and left several firefighters struggling for their lives; after the Northridge earthquake and a litany of other disasters in the 1990s; it's little wonder that many firefighters and officers reacted with wary bemusement when prominent geologists issued public warnings about yet another threat: mud and debris flows. These flows were expected after devastating wildland fires raked southern California's mountain slopes, leaving thick layers of ash and destabilizing the soils and boulders by burning out the root systems.

The general response among some fire officials was a resigned, "How much time do we have to prepare for *that* disaster?"

The warnings about mud and debris flows first came to prominence on the heels of the firestorms. Within days of the wildfires that struck in October 1993, local geologists and forestry officials expressed concern that heavy rainfall would cause scorched hillsides to slough off in thick layers of ash and mud, only to be followed by large boulders, automobiles, homes, and people. As all of this material poured into the bottoms of canyons and creeks, they warned, it would collect, mix, and eventually become mud and debris flows capable burying entire neighborhoods when rain fell in sufficient amounts to mobilize it.

When a series of mud and debris flows roared from the mountains above Los Angeles during winter storms, homes were filled to their roofs with families trapped inside, some people having floated up to the ceiling, still alive, as cars washed away and entire neighborhoods were wiped out. The hillsides had been denuded by massive wildfires the previous fall, setting the stage for further disaster when the rains came. L.A. County firefighters rescued

some people by tunneling through roofs and down through the ceilings to reach them (Fig. 2–6). Events like this defy most mitigation measures. Mud and debris flows challenge incident commanders to conduct organized, methodical, yet timely operations under very hazardous conditions to save lives.

Fig. 2–6 L.A. County Firefighters Tunnel through Roofs to Rescue Trapped Victims

In this case, the term *mobilize* is a relatively innocuous term to describe what really happens. Mud, debris, and water are suddenly unleashed in a sort of flash flood, capable of scouring entire canyons and wiping out anything in its path. Mud and debris flows are typically preceded by a leading edge that consists of boulders, trees, homes, cars, bridges, and other objects that get swept into the flow. For victims in the path of a mud and debris flow, the effect is not unlike facing an oncoming avalanche of mud that can travel at more than 100 miles per hour. These storms can strike with devastating fury, penetrate everywhere, and bury entire neighborhoods.

Fortunately, the disaster-weary fire/rescue services, supported by public works agencies and flood-control districts, heeded the warnings and embarked on a campaign of mitigation, training, and other tightly focused preparedness initiatives to address the consequences of mud and debris flows. Fire department swift-water rescue response matrixes were adjusted to include the response of more first alarm resources. Mud diversion barriers were installed in vulnerable neighborhoods, debris basins were cleared, and firefighters were trained to conduct mud and debris flow rescue operations. These rescue operations differ in key ways from typical swift-water rescues because of the severe life hazards created by the suspended debris in the mudflows. As each major storm approached, heavy equipment, helicopters, USAR companies, and swift-water rescue teams were deployed to strategic locations in vulnerable areas.

True to the warnings, southern California experienced a series of deadly mud and debris flows that—despite preparedness efforts—wiped out entire neighborhoods, buried hikers beneath thousands of tons of mud and debris, and swept little children from the arms of their parents as homes disintegrated beneath floods of mud and stones. Bad as it was, it could have been much worse. If not for the advanced levels of preparedness among local fire/rescue agencies, public works, flood control districts, and the general population, the toll in lives and property loss across southern California would have been greater.

The mud and debris flow story is representative of effective disaster preparedness. It's a demonstration of how disaster threats can be recognized, quantified, and addressed if local decision-makers are inclined to realistically view conditions and trends, and if they are prepared to address potential scenarios in a straightforward manner. With effective planning, our cities and towns are not predestined to suffer maximum tolls every time a disaster strikes. Even the most destructive disasters can often be mitigated through pro-active, thorough, and well-considered assessment and preparedness efforts.

Following the destruction of the Alfred P. Murrah Building on April 19, 1995, most people suspected that the act was a result of international terrorism. Within days however, the nation and much of the world was shocked to learn that the real perpetrator might have been a former U.S. soldier named Timothy McVeigh.

As the emerging threat of international terrorism continued to evolve, efforts to plan, train, and prepare were mixed. Although it should have been clear that the first responders of local fire/rescue, police, and EMS agencies are the nation's first line of defense against the consequences of successful terrorist attacks, relatively little funding was designated to prepare them for the problems they might likely encounter. Instead, the majority of *consequence management* funding went to agencies whose forces were certain to arrive many hours (or in some cases, days) after a terrorist attack, leaving the first responders vulnerable to the lethal effects of terrorism. The terrorist attacks on September 11, 2001 crystallized the dilemma faced by first-arriving firefighters, USAR teams, police, and EMS personnel.

In contrast to those more innocent times before the tragedy in New York, today it should be assumed that every major city may be a target of terrorism. Indeed, even smaller cities and towns should consider the potential for terrorism of one form or another. Therefore, cities and towns both large and small, as well as the states themselves, should have adequate DROPs in place. The new paradigm of terrorism is a dramatic example of how an emerging threat necessarily changes the process of disaster planning and preparation.

Case study 2: the threat of near-source tsunamis

 In the world of disaster, there are not-so-obvious examples of emerging disaster threats that must be taken seriously by fire/rescue authorities and disaster planners. For instance, as of this writing, fire department officers and disaster planners in southern California coastal communities are developing DROPs to manage the consequences of tsunamis up to 30 ft high (or higher). These tsunamis could strike the coastline within minutes of a local thrust-fault earthquake or an underwater landslide, wiping out large sections of heavily populated coastal communities and the ports of Long Beach and Los Angeles.[7] For many years, most scientists considered the potential for *near source* tsunamis along this part of the Pacific coast nearly nonexistent. However, new data has demonstrated that locally generated tsunamis can occur (and already have occurred in the past) in some of the most densely populated areas.

What does this mean for fire and rescue personnel and the public they protect? The answer is simple. In post-earthquake operations where fire/rescue resources may be deployed unknowingly into potential *tsunami impact zones* to fight fires, conduct SAR in collapsed buildings, mitigate hazmat problems, and treat mass casualties, the sudden and unannounced arrival of 10-meter tsunamis capable of rushing nearly one mile inland could have deadly consequences for both victims and rescuers.

The new revelations about tsunami hazards are prompting fire/rescue officials in southern California to revise existing earthquake DROPs and develop new tsunami disaster plans. In a sort of real-time test of the concepts established in this chapter, this author and a number of fire/rescue personnel, lifeguards, law enforcement officials, emergency planners, and civil authorities in L.A. County are helping to develop disaster plans to address these concerns. These plans will provide policies, procedures, and guidelines to help protect the public and the public safety agencies that respond and are affected by near-source tsunamis, particularly those coming fast on the heels of a damaging earthquake.

The first four steps of disaster planning have been completed. The agencies with primary responsibility for rescue operations (the fire department and lifeguard agencies) have been identified. Hazard assessments, including computer-generated hazard studies conducted by an internationally recognized tsunami expert from the USC School of Engineering, under contract to the California OES and FEMA, are nearing completion. Rescue resources have been inventoried. Step 4, writing a county-wide tsunami disaster plan, as well as writing tsunami disaster plans for each affected fire/rescue agency, is in the final stage. If one or more near-source tsunamis strike southern California in the coming years, the efficacy of the disaster planning process will be severely tested.

Measuring Disaster Potential: Looking to History

When assessing the potential for rescue-related disasters and planning effective responses, it's sometimes instructional to look at history for examples of what can happen, how disasters can affect society, and the impact that such events have on fire/rescue professionals and others whose job it is to protect the public. We are so often caught up in the events of the day that we tend to ignore the overall context in which they occur. This is especially true of disasters, which may recur infrequently in a given area. Because our experience with them may be limited, disasters are sometimes difficult to place into their historical context. As firefighters and rescuers, we have a responsibility to address the consequences of disasters that occur *now*. But it's often helpful in the planning and preparedness process to consider the larger picture, to understand how disasters may affect our communities in the future.

A good example of the difficulty of determining a historical context for disasters was noted in a review of the process of planning for near-source tsunamis in southern California, discussed previously in this chapter. When tsunami experts began warning that evidence had been discovered indicating the very real potential for seismic sea waves to strike the coastal zones following an earthquake or underwater landslide, one of the first questions raised was: How often do these events really happen? Another question was: Is there historical evidence of near-source tsunamis?

These questions were difficult to answer because the coastal zones of southern California have been visited and populated by the Spanish and Europeans only for a few hundred years, and because the Chumash and other Native American tribes didn't leave clear records indicating that tsunamis had struck their villages and places of habitation.

Historic reminder of the importance of disaster planning

Throughout history, tsunamis, earthquakes, and volcanoes have been archetypes of disaster around the world. There was never a time when these forces of nature weren't a factor in human survival somewhere in the world. Movement of the earth below our feet, mountainous ocean waves, and mountains spewing molten rock are among the oldest and deepest fears in human consciousness, and for good reason. Entire civilizations have been altered or destroyed by such events. When these events occur in modern times, fire and rescue personnel are expected to have a plan and to react accordingly. Disasters on various scales will continue to occur, and in many cases the effects are worse because more people are living in towns and cities vulnerable to disaster. There has never been a time when firefighters and rescuers faced more serious challenges from disasters around the world.

The modern Greek island of Santorini is a remnant of the Aegean Sea volcano known as Thera, which blew itself off the face of the earth around 1600 B.C. A similar event is certain to be repeated in various parts of the world in the coming years or decades. The island of Thera was the site of one of the largest volcanic eruptions in history. Fortunately for the local residents at the time, the island volcano forewarned local populations of the fury about to be unleashed. Several weeks before the island was vaporized, the region was jolted by moderate earthquakes and Thera experienced a relatively small eruption. The events were apparently of sufficient force to frighten some people away from the island. One might speculate that the population of Thera exercised an evacuation plan of sorts, and none too soon!

The warnings continued as the volcano grew more dangerous. The entire region continued to be rocked by earthquakes. Thera's next eruption was more dramatic, burying the Minoan city of Akrotiri deep in volcanic ash. Scientists speculate that it was this eruption that caused the evacuation of the Minoans from Thera. Their speculation is supported by excavations of the layer of volcanic ash representing the largest eruption. There is little evidence of people dying on Thera.

For many, however, evacuating to another island made little difference, because a far more catastrophic event was about to occur. Within weeks of the most serious foreshocks, Thera suddenly came to life and erupted with a force that caused most of the island to disappear, its mass blasted skyward as the volcano vented its contents into the atmosphere. With most of its mass suddenly ejected, the edges of the volcano collapsed into the center. This was to have catastrophic consequences. The Aegean Sea suddenly began pouring into the vent of the collapsing volcano.[8] As seawater whirlpooled into the throat of the incandescent caldera, it was instantly vaporized, expanding into steam and causing an explosive eruption heard as far away as Gibraltar, Scandinavia, the Arabian Sea and Central Africa.

It's been estimated by some researchers that Thera's eruption was loud enough to deafen people within several hundred miles. The sudden release of energy caused tsunamis estimated as high as 150 ft to race across the Aegean Sea, devastating the shores of many populated islands. Thera's eruption left an 80-square-mile caldera more than 1,000 ft deep beneath the turquoise-blue Aegean Sea. The caldera was—and today remains—surrounded by a series of islands arranged roughly in a semicircle marking the outlines of the once-great volcano.

The Santorini eruption apparently was catastrophic to the highly advanced Mycenaean civilization that had taken root on Thera, Crete, and nearby islands.

Archeologists working on Santorini recently discovered the remains of palaces and other large buildings buried in thick layers of volcanic ash, complete with frescos and decorated pottery intact, just as it was when buried. The ruins have been dubbed the *Pompeii of the Bronze Age.*

Scholar Angelos Galanopoulos considered the eruption of Thera a turning point in human history. In describing the event, Galanopoulos wrote:

The collapse of Thera was a disaster comparable to nuclear war today. Hundreds of thousands of persons could have lost their lives. Cities, ports, and villages on many islands and on the mainland of Greece and Turkey could have been washed away or inundated by torrential rains triggered by a spew of ash. What remained would have been toppled and pounded to rubble by tidal waves. Fleets of ships would have foundered or been hurled miles inland. Cities on the highlands, at least those close to Thera, would have been rocked and torn by earthquakes. And all the while, volcanic ash would have blackened the heavens, turning day into night, with thunder crashing, lightening searing the sky, and the seas becoming clogged with mud. [9]

As early as 1939, it was suggested that the highly cultured Minoan villages of Crete, located across the Aegean Sea from Thera, had been destroyed not by wars or political upheaval, but rather by devastating earthquakes. One quake in A.D. 365 uplifted parts of the island of Crete by 27 ft. Giant tsunamis and aerial vibrations followed Thera's explosive eruption and collapse.

For 1,500 years, beginning about 3000 B.C., Crete and the Cyclades dominated the Mediterranean. Here indeed was the birthplace of aspects of later European civilization. But the great society failed for reasons that for centuries remained an enigma. Then archeologists began discovering tantalizing clues to the disappearance of the Minoans. After discovering the ruins of Minoan palaces and sophisticated murals on Crete, one researcher wrote, "[W]hat really piqued my interest...were the curious positions of several huge stone blocks that had been torn from their foundations and strewn toward the sea." The researcher went on to say, "I found a building near the shore with its basement full of pumice. This fact I tentatively ascribed to a huge eruption of Thera." And so the mounting evidence pointed to a natural calamity of huge proportions as the reason for the disappearance of the highly advanced Minoans.

So great a disaster was Thera's collapse and associated destruction, that it left an indelible impression on human culture. The recurring human and religious theme of great floods, which according to Biblical scripture, caused the destruction of highly evolved societies in various parts of the world, has been characterized by some experts as shared human remembrances of calamities that actually happened. These remembrances were recalled through oral and written histories, stories about calamity that have been passed from generation to generation, the facts sometimes becoming distorted during repeated tellings. And so it was with Thera. In his 1960 paper entitled, "Tsunamis Observed on the Coasts of Greece from Antiquity to the Present Time," Galanopoulos attributed the eruption of Thera as the basis for Deucalion's flood, the Biblical deluge, the plagues of Egypt, and other reported catastrophes that have left indelible marks on Western civilization.

Disaster plans should take long-term operations into account (Fig. 2–7).

Fig. 2–7 Modern Manmade Disaster

There are of course many examples of earthquakes, volcanoes, and other disasters changing the course of human history. Consider Pompeii, the highly cultured Roman city buried beneath superheated ash and debris from the infamous Mt. Vesuvius, which erupted in 69 A.D., leaving a virtual freeze-frame of life in those times. And think about the destruction of Lisbon in the 15th century, yet another reminder of the history-altering effects of earthquakes.

Case study 3: great flood and other disasters

Sometimes mythical or even Biblical disasters hold lessons for modern fire/rescue professionals in terms of planning for unusual events that can overwhelm existing response systems. Take for example the unlikely lesson of Noah's flood. Some researchers are now convinced that the motif for the great flood may have originated with an actual catastrophe that occurred in modern-day Turkey. The Sea of Marmara, which borders the historic city of Istanbul and the Turkish and Greek coasts, is actually a submerged rift zone nearly 4,000 ft deep, where the land has been pulled apart by a split in the North Anatolian Fault and filled in by water from the Mediterranean Sea.[10] At the end of the Ice Age some 12,000 years ago, the Sea of Marmara was completely separated from the freshwater Black Sea by a narrow bridge of land in what is now the Bosporus Strait.

The Sea of Marmara was at that time several hundred ft higher in elevation than the Black Sea. Imagine a large, high bowl of water next to a small, low bowl of water. That would be the Sea of Marmara in relation to the Black Sea. As the continental glaciers melted, the Marmara Sea rose, threatening to spill over the natural dam of the Bosporus, where it would pour into the Black Sea. Some researchers speculate that an earthquake along the North Anatolian fault caused the Sea of Marmara to slosh across the Bosporus, cutting a channel that became a flume of water flooding down into the Black Sea, eroding away the land bridge until it ruptured. The ensuing flood roared for more than 100 days as the Sea of Marmara poured its contents into the Black Sea, submerging towns and villages under a flood that spread inland more than a mile each day and sending the population fleeing to higher ground. Scientists believe that the roar of the flood may have been heard as far away as China.

The flooding continued until the Black Sea basin reached sea level, and melded with the Black Sea. Today the Bosporus Straight marks the place where the two seas met, and large ships now sail from the Mediterranean through the Sea of Marmara directly into the Black Sea via the Bosporus.

It's true that entire civilizations may already have been erased from the face of the earth by earthquakes, tsunamis, and volcanic eruptions. What's to prevent it from happening again, and how can we reduce the societal impact of such an event to the lowest level possible? In recent years, the eruption of Mt. St. Helens in the state of Washington, a Congolese volcano, several devastating eruptions in Colombia and Costa Rica, and an eruption at Mt. Pinatubo (Philippines) have demonstrated the need for rapid response to conduct search, rescue, and evacuation operations. The long-term forecast for eruptions of dangerous volcanoes near Mexico City, Seattle, Naples, Pompeii, and other major cities makes it clear that the fire/rescue services will continue to confront challenging disaster rescue operations resulting from volcanoes.

In the greater Seattle area alone, major efforts are underway to address the consequences of the eruption of Mt. Rainier, which is likely to melt the snowcap and unleash deadly mud and debris flows down its slopes toward the cities. California officials are updating plans for escape, evacuation, and SAR operations in the Mammoth lakes area of the Sierra Nevada mountain range in preparation for an eruption that threatens to cut off escape, trap residents, and even bury parts of the aqueduct system. The fire and rescue services will continue to play a major role in these operations in the event of actual eruptions.

Assessing the Potential for Earthquake Disasters

Earthquakes are clearly one of the most common disaster scenarios for large-scale urban SAR operations. In many places, earthquakes are the primary reason for the development of USAR response systems in the first place. And although terrorism and other events have caused massive structure collapses and consequent USAR operations that have lasted for weeks, earthquakes continue as a potential source of catastrophe in the form of hundreds or even thousands of simultaneous structural collapses that can overwhelm the rescue resources of practically any nation.

In the San Gabriel Mountains above Los Angeles, the signs of earthquake potential are all around. The San Gabriel Fault clearly shows faulting that has caused large earthquakes and built entire mountain ranges in southern California over tens of thousands of years. The 1971 Sylmar earthquake (7.1 on the Richter scale and 65 fatalities in Los Angles County) was caused by a fault like this. Several large dams have been built within 15 miles of this fault with hundreds of thousands of people living downstream. Earthquake disaster plans should take into account potential secondary consequences like dam failure, which will complicate post-earthquake fire-fighting and USAR operations and in some cases may kill more people than collapsed buildings (Fig. 2–8). The Northridge quake occurred before dawn. What would be the consequences of the same earthquake during business hours and with this venue full of people?

Fig. 2–8 A Sign of Disaster Potential

To facilitate the development of disaster plans and rescue systems that will prove effective when damaging or even catastrophic earthquakes strike, fire chiefs, rescue companies, emergency planners, and other decision-makers should have a basic understanding of local seismic hazards. They should know the location, frequency, and magnitude of earthquakes that may be expected to result from the rupture of local faults. Just as a firefighter should understand basic fire science to accurately anticipate (and respond to) the behavior of fire, so should the fire/rescue professional understand enough seismology to link ground motion with possible damage to the community. This knowledge will make the fire service respondent to an earthquake better prepared to manage the consequences.

The 1994 Northridge and 1995 Kobe earthquakes are examples that demonstrate the acute need to develop closer working relationships to improve information sharing between seismologists, fire/rescue professionals, and emergency planners. This is especially evident when one considers the rate at which new seismic hazards are being discovered.

The Northridge earthquake occurred on a buried thrust fault whose presence was not even known until it ruptured nine miles beneath the surface of the San Fernando Valley. In the wake of Northridge, seismologists have discovered (or postulated) the presence of many more buried thrust faults—some of which appear to be capable of delivering even more devastating blows to densely populated areas—beneath the surface of greater Los Angeles. It's now recognized that hidden thrust faults are a major cause of the tortured topography found in parts of southern California and may represent nearly as great a danger to the large cities as other well-known features like the San Andreas Fault.

And the lessons keep coming as more research focuses on assessing the potential for earthquake disasters and as progressive fire departments increase their interaction with the scientific community.

The Importance of Cooperative Efforts

Sources of important seismological information necessary to develop DROPs may include local universities and geological or seismic research centers, state geological and seismic agencies, and the National Geological Survey. It's increasingly important for fire/rescue agencies to be familiar with this information, and to interact with the people whose research brings the information to us. Pro-active fire/rescue agencies in seismically active areas derive huge advantages from collaborations with the organizations, schools, and agencies whose mission it is to study the earth and to provide information on earthquakes to the public and to public safety agencies (Fig. 2–9).

On the part of firefighters, rescue teams, chief officers, and emergency planners, this collaboration is leading to a better understanding of the forces that drive the seismic disasters that periodically strike, and the consequent development of more effective earthquake disaster plans. It is a relatively new partnership that will enhance the planning process for earthquakes that are certain to strike the region. The old adage, *know your enemy* (in this case the enemy being the consequences of ground fault rupture), is a good analogy here. The more firefighters and rescuers know about earthquakes and the hazards they bring, the better prepared they will be to manage the consequences.

Fig. 2–9 Damage from the 1994 Northridge Earthquake

Knowing the probability of future seismic events and their possible magnitude has helped seismologists better understand the needs of emergency responders. As a result, advanced warning systems and rapid identification of the location and magnitude of ground fault ruptures helps responders to rapidly deploy resources. We are finding that there are ways in which seismologists, other earth scientists, and researchers can enhance the ability of fire/rescue agencies to react to earthquakes.

Case study 4: an ancient lesson in the consequences of disaster

How can we relate past disasters to planning for disasters in the modern world? One example is found in the fate of two old world cities known as Thonis and Menouthis, in Egypt. The events that destroyed these ancient cities may serve as a parable for the dangers that confront many modern-day cities like Seattle, Anchorage, Istanbul, and other places where people live along coastlines subject to sudden land subsidence from earthquakes and/or volcanoes.

Thonis, also known as Herakleion in the ancient world, was the largest port city serving the Nile until Alexander the Great built his namesake port of Alexandria in 331 B.C. Although Thonis's renown as a port declined, it remained a bustling city for several centuries. Menouthis, an elegant city built a few miles inland of Thonis, was an important center to several religions of the day. It was the site of a temple dedicated to Isis built in the 4th century B.C. Tens of thousands of people flocked to Menouthis to pay homage to their particular god or gods.

Both cities vanished sometime between the 1st and 4th centuries A.D., and the cause of their demise was a source of speculation for hundreds of years. The very existence of Thonis and Menouthis was disputed by scientists for many years because no physical evidence had ever been found. But in 1999, a French research team exploring the waters around Aboukir Bay near Alexandria found the tumbled remains of both Thonis and Menouthis beneath 30 ft of seawater, right where the Mediterranean swallowed them up more than 2,000 years ago. Not only are the cities virtually intact, frozen in time since the day they were submerged, but physical clues point to great earthquakes, tsunamis, or some other force of nature that first leveled, then submerged, both cities.

Virtually all the ruins of Thonis lie tilted in the same direction beneath the water, a footprint of some giant force that knocked them down and sank them. The tilting effect is not unlike a house of cards built on a tablecloth. If one pulls the tablecloth out from beneath the house, the cards tend to fall in the same direction. It's clear that the buildings of Thonis weren't destroyed by post-submersion wave action or underwater currents. Instead, some scientists are certain they fell during a great earthquake, and then the land upon which the city was built sank below the sea, perhaps in the span of a single day. Other researchers speculate that Thonis was destroyed by a tsunami generated by a giant earthquake elsewhere in the Mediterranean Sea or perhaps from a volcanic eruption. Some consider the possibility that a large earthquake and series of tsunamis wracked the city on the same day.

After 400 years, another very large quake and perhaps a series of tsunamis may have hammered Menouthis and submerged it beneath the surging waves. The evidence for a tsunami is growing as researchers find more indications that the Mediterranean is the site of a large number of devastating waves that have swept across its breadth throughout (and before) recorded history. One doesn't have to go back very far to find evidence of the power of tsunamis in the region. Lisbon was devastated by a tsunami and earthquake that killed more than 60,000 people in 1755. Today, modern Lisbon and many other Mediterranean cities remain under threat of a similar fate.

What, one might ask, does this extended discussion of ancient historical catastrophes have to do with planning for modern urban search and technical rescue disasters? As often happens with other subjects, there is a common thread that runs through history and modern reality with regard to disasters.

It turns out that the ancient submerged cities of Thonis and Menouthis have something in common with modern cities like Seattle, Anchorage, and certain coastal cities built in subsidence zones in Turkey, Indonesia, Japan, Taiwan, South America, and other parts of the world. It has become evident that parts of these cities, nations, and continents are possible candidates for the fates shared by Thonis and Menouthis. It's postulated by many experts—based on studies of history, geology, volcanology, and seismology—that parts of each of these places may someday disappear from sight, swallowed up by the oceans during catastrophic earthquakes. It's happened before in these places.

People living today in cities like Seattle and Anchorage and Marina Del Rey (California) may have difficulty imagining Main Street laying in a watery grave beneath 30 ft of water for the next several thousand years. In fact, places like Marina Del Rey are more likely to be swamped by tsunamis rather than sinking beneath the Pacific, but may still find themselves under water at least temporarily from the results of earthquakes or other major disasters. Surely, the good citizens of Thonis and Menouthis never anticipated that the only visitors to their fair cities for the next two millennia would be sea anemones and French-speaking scuba divers.

Could something like that really happen today in a populous coastal region of the U.S. and other modern nations? The answer has to be yes. In fact, it recently happened when Turkey was hammered by a devastating series of earthquakes in 1999. During the first quake at 3:00 a.m. on August 17, some coastal sections along the Gulf of Izmit (Turkey) dropped nearly 30 ft, immediately inundating waterfront neighborhoods,

the rush of seawater meeting a shock wave of dust from collapsing buildings along the coast.

Many buildings collapsed even as they went beneath the water. Residents of many buildings sank beneath the Sea of Marmara with their homes, unable to escape. They were not unlike the crew of a submarine or a ship being trapped within a sinking vessel. Two police officers sitting in their squad car on a residential street drowned when their vehicle suddenly plunged beneath the waves with the rest of the neighborhood. It took dive teams weeks to search submerged cars and cafes for the bodies of those who were drowned with the buildings and streets. The coastline itself was altered forever, and some modern buildings are now under water. The Izmit disaster is proof that even large modern coastal cities can go the way of Thonis and Menouthis.

Can anything be done to prevent entire coastal zones from submerging when the earth's crust ruptures? Obviously the answer is no. Can effective measures be taken to limit the life loss in the subsidence zones? Perhaps. Will proper assessment of the hazards and pro-active response help firefighters and USAR teams save lives in areas bordering the inundated zones? Probably. Can awareness of such unseen hazards and intelligent planning prevent unnecessary loss of life among emergency responders? Certainly. Prevention starts with awareness based on proper assessment of disaster potential.

Structural Design and Construction Issues

In quake-prone areas, public safety relies on effective building codes, which are in turn based largely on the ability of the engineering community to design buildings to resist the effects of violent seismic shaking. When structures fail during earthquakes, firefighters and rescue team members are responsible for managing the consequences, which may include fire, hazardous materials releases, and live victims trapped within highly unstable buildings and rubble piles.

Despite remarkable advances in earthquake-resistant engineering and retrofitting, buildings continue to fail in varying measures during large quakes. Millions of existing buildings and other large structures (i.e. freeway overpasses, tunnels, etc.) were built before the advent of modern seismic codes and are not subject to retrofitting campaigns, even when active faults are found nearby. This is true of many regions of the world, and it's equally true of the United States, including the New Madrid fault zone, New York City, and other places that were not traditionally considered quake-prone.

In California, long known to be vulnerable to earthquakes, the government released a report in March 2001 listing 966 of the state's 2,467 hospital buildings as posing "a significant risk of collapse and a danger to the public after a strong earthquake."[11] In L.A. County alone, the report said, more than half the hospitals surveyed risk collapse in a strong earthquake and must be retrofitted or removed.

Reports like this include the kind of information that should garner the immediate attention of fire chiefs, station captains, tailboard firefighters, and USAR teams alike. Hospitals are of significant importance, not only because they are the backbone of a community's medical system, but also because they usually represent serious life threats because of their size and occupancy.

The collapse of a single hospital presents a severe threat to the welfare of the community that it serves. The sudden reduction of available medical services has a serious impact on the public and also on adjacent hospitals that must deal with the overflow patients who would normally stream into the collapsed hospital. In addition, hundreds (if not thousands) of doctors, nurses, and other caregivers who would normally be assisting with disaster operations may themselves become trapped victims. For those whose first-in districts include major hospitals subject to collapse in

large quakes, it's a problem of the first magnitude. For fire/rescue agencies whose jurisdictions include hospitals and other critical facilities, it's only prudent to have a plan to deal with such an eventuality.

Many other non-quake-hardened structures in earthquake-prone regions are vulnerable to catastrophic collapse, even during moderate or minor earthquakes. Despite the remarkable performances of modern quake-resistant structures, there are cases in which these newer buildings fail during large earthquakes. Consequently, quake-induced structural failures (often followed by secondary collapse during aftershocks) will be a part of life in major metropolitan areas around the world in the coming years. Not only will collapsing structures threaten the inhabitants, but they may also exact a toll on firefighters and other rescuers called upon to locate, treat, and extract victims trapped within them.

This is not to downplay the effectiveness of new earthquake-resistant designs in reducing building failures. To the contrary, evidence from recent earthquakes indicates that new and retrofitted structures are safer than ever. Still, even quake-resistant structures continue to fail in significant earthquakes. When they do, some of the most valuable lessons are learned by the emergency responders whose job it is to deal with the consequences. Given this fact, there is a need for greater interaction between seismologists, the engineering community, and emergency responders to increase the survivability of citizens who live and work in these structures and to improve the safety and effectiveness of post-quake USAR operations.

Many of our cities have hundreds of miles of underground pipelines carrying hazardous, flammable, and explosive materials, any of which could rupture during the next earthquake. Water mains may rupture, as they did in the Northridge earthquake, creating big-time problems for firefighters across L.A. County. Few people can conceive of the possible effects of hundreds of fires starting simultaneously in many cities across a large band of earthquake-impacted land. These fires can spread to damaged buildings in which victims are trapped, overwhelming firefighters who must decide between taking the time to rescue victims or trying to knock down the fire before it gets to them. Even fewer can conceive of such an event occurring on a hot, dry day with winds whipping across densely packed urban areas, spreading flames on a scale previously seen only in wildland fires.

Add to that the potential for dam failure, which according to studies by the U.S. Geological Survey, could kill thousands of people in quake-prone places where cities, towns, and settlements are located downstream of large dams. What if there was another simultaneous disaster, upon which the effects of the earthquake would be *layered*? The possibility could even exist for a series of very large tsunamis to wipe out extensive areas of densely populated coastline just minutes after another big quake. Scientists at USC are currently preparing computer models of the impact areas, and various agencies are developing public education programs and response plans for just such an event.

It's disconcerting that many public officials continue to ignore plausible warnings about earthquakes. Certainly it's past time for an awakening to the fact that future earthquakes in the United States could cause destruction and death on a par equal to that seen in Turkey, Taiwan, Kobe, Soviet Armenia, Mexico City, and other places in recent years.

Other Common Disasters

Some disasters are universally feared because they happen on practically every continent on earth (or in the world's oceans). The list of potential disasters is practically endless, and it expands as society, technology, and environmental factors change. Here is a brief sampling of other common disasters for which some fire/rescue agencies should be prepared.

Dam failure

Dam collapse is among the most lethal singular events that plague society. For example, one of the most notorious disasters in U.S. history was a collapse that occurred in 1928 in L.A. County, resulting in more than 500 deaths and second only to the 1906 San Francisco earthquake in lethality. At 11:57 p.m. on March 12, a geologic defect caused a giant landslide that crushed part of the 1,300-ft wide St. Francis Dam. The dam failed and sent a 90-ft high, 12 billion gallon wall of water roaring through the Santa Clarita Valley, an area served by the LACoFD some 30 miles northwest of downtown Los Angeles. A woman whose home was perched in a deep canyon downstream of the dam later recalled bolting awake at the indescribable sound of something coming in the darkness. She reported "a haze over everything" and instantly knew that the dam had failed. Her husband and two daughters perished when they were unable to get out of the house in time.[12]

On its way through a river valley to the Pacific Ocean, 54 miles away, the flood destroyed more than 1,200 homes and 10 bridges and wiped several towns practically off the map. Although many deaths occurred in the first 30 minutes, it took $5^1/_2$ hours for the flow to reach the Pacific Ocean. As frantic word spread that the dam had collapsed, telephone operators in towns in the path of the flood called ahead to warn others to evacuate. The first official warning was issued at 1:20 a.m. after the flood traveled through 18 miles of the Santa Clara River Valley, devastating towns along the way.

In a camp established to construct an aqueduct, 84 of 150 workers were killed as the flood swept past. Some people survived by escaping to high ground or by riding floating debris in the inundation. One man survived as the water washed over his Chrysler while he was trying to escape, but a passenger with him attempted to climb out of the car and was never seen again. A California highway patrolman rode his motorcycle ahead of the flood, warning towns located downstream that a wall of water was coming. His actions saved hundreds of lives by warning sleeping residents to escape to high ground.

Bodies washed up on beaches from Ventura to San Diego. SAR operations required weeks of difficult work by firefighters and sheriffs' deputies who had no swift-water rescue teams or rescue companies on which to rely. They had no helicopters and no USAR teams. Instead, rescuers were forced to rely on heavy equipment and all-terrain vehicles to operate in deep mud and debris, extract victims from debris piles and hardening mud, and transport the injured and the dead.

Today, the Bouquet Reservoir dam holds back billions of gallons of water in the canyon one ridge over from the old San Franscisquito Dam site. Given the vastly increased population in the areas devastated by the 1928 collapse, experts now predict that more than 20,000 fatalities might occur if the Bouquet Reservoir dam ruptured from an earthquake, flood, landslide, simple structural failure, or terrorist attack. A number of other dams in L.A. County could cause tens of thousands of deaths if they collapsed.

The worst failure of a man-made dam occurred in Henan Province, China, resulting in more than 230,000 deaths. In August 1975, the Banqiao and Shimantan dams collapsed almost simultaneously from geological problems and structural weakness.[13] Could something like that happen in the United States? Clearly, this should be an issue of concern for those charged with responsibility for ensuring preparedness to manage SAR operations after such an event, wherever it may occur.

Train disasters

The worst train disaster as of this writing occurred when a train plummeted off a bridge into the Bagnati River in India in 1981. Between 800 and 900 people died. Today, commuter and freight trains race through our densely populated urban areas 24 hours a day. They travel through long tunnels, across rugged terrain, and under unfavorable weather conditions that can undermine the tracks. They often carry lethal cargos of toxic chemicals.

Dozens of people every year are killed in train mishaps across the United States. Passenger train derailments can challenge firefighters and rescuers with multiple casualties and complex rescue scenarios (Fig. 2–10). The potential for a truly catastrophic train derailment or collision is certainly there. Is your community prepared to manage the consequences of such an event?

Fig. 2–10 Derailment in Burbank, California

Underground disasters

On February 15, 2003, a man lit a can of lighter fluid and tossed it into a Seoul (South Korea) subway train, igniting a fire that quickly burned the contents and trapped hundreds of people in the train. Moments later, another train pulled into the subway station and caught fire, trapping even more people. Hundreds of people died in the inferno, which created a nightmare scenario for firefighters attempting to locate and control the fire, search for victims, and extract live victims. It was an indication of just how easy it is for terrorists (even those operating alone) to cause havoc and death in enclosed public spaces beneath the ground.

In 1995, a passenger train caught fire in a tunnel beneath the Azerbaijani city of Baku, killing 300 people and injuring 250. Every year dozens (if not hundreds) of miners die in accidents in mines in North America and other continents. Underground disasters require specialized equipment, highly trained teams, and a level of expertise that's difficult to find in the typical city, region, or state.

Road tunnels

In 1982, more than 170 people died when a gas tanker caught fire and blew up inside the Salang Tunnel in Afghanistan. In the Swiss and French Alps, hundreds of people have died in similar accidents in road tunnels. What are the requirements to manage the consequences of such a disaster?

Ski lifts

In 1976, in the Italian resort of Cavalese, 42 people died when a cable lift ruptured. Dozens died when the cable supporting an Italian ski lift was struck by an American jetfighter. In Lake Tahoe, a number of people were killed and dozens trapped when the cable cut through a car suspended hundreds of ft above the rocky terrain. Is your department ready to manage such a disaster involving many victims trapped above ground?

Airplane disasters

Except for the terrorist attack on the World Trade Center and Pentagon (Fig. 2–11), the worst loss of life from an airliner disaster occurred on the ground in the Canary Islands. Two airliners collided in fog, killing 583. In 1985, a Japan Airlines Boeing 747 crashed, killing more than 520 people. In 1996 351 people died when a Saudi Boeing 747 collided in midair with a charter airliner over New Delhi, India. In Sioux City, Iowa, a commercial jetliner cartwheeled down a runway with fire/rescue units standing by because the plane lost its hydraulic control systems. It could happen at practically any commercial airport. Is your department ready for such a rescue disaster?

Sep. 12, 2001, 17:37:23   #4 impact

Fig. 2–11 Airliner Striking the Pentagon on September 11, 2001

Mountain rescue disasters

In July 1990, 43 climbers were buried and killed by avalanches that swept down the face of a mountain in Tajikistan. Every year people are buried by avalanches in North American mountain ranges. Is it possible for your department to be called upon to assist in SAR operations in such a mountain disaster?

Structure collapse

During the reign of Antoninus Pius (138-161 A.D.), more than 1,100 spectators died in the collapse of the upper decks of Circus Maximus in Rome, Italy. The collapse occurred in the middle of a gladiatorial combat. The 1998 earthquake in Turkey killed between 17,000 and 45,000 people. In 1918 the Hong Kong Jockey Club collapsed and burned, killing more than 600 people. One earthquake in China reportedly killed 250,000 people. A shopping mall in Seoul (South Korea) collapsed, killing hundreds and trapping victims alive for 16 days. The largest and most deadly single building collapse occurred during the 9-11 attacks when the Twin Towers burned and fell. Is your department ready for a major collapse rescue disaster that traps dozens, hundreds, or even thousands of people alive?

The Northridge earthquake (fortunately) occurred during early morning hours when the stadium was deserted. Consider the challenge of rescuing dozens of people trapped beneath a collapse of this nature, with a crowd of tens of thousands of people present (Fig. 2–12).

Fig. 2–12 Aftermath of Northridge Earthquake

Public works disasters

In 1970, construction on a water tunnel was begun to bring water to L.A. from the Feather River project in central California. The tunnel was to pass directly through several earthquake faults on its way beneath the mountains of the Angeles National Forest to the northern desert. Just months after the 1971 Sylmar earthquake caused 65 deaths in the north San Fernando Valley, a giant explosion killed 17 tunnel workers as they bored through a section of the tunnel that was within one-half mile of the fault that had caused the quake. This multi-day rescue and firefighting operation required firefighters to use cranes, tunneling equipment, and a host of unusual tactics.

In recent years, rescue mishaps beneath and within urban Los Angeles have confronted firefighters with similar challenges. These included the Metro Rail tunnel fire in 1990 and the infamous sinkhole collapse, an event that came within seconds of killing up to twenty firefighters and construction workers in the twin tunnels that run 80 ft below Hollywood Boulevard.

Pipeline disasters

Buried pipelines are an example of unseen disasters waiting to be sprung on modern society. Today there are more than 165,000 miles of underground pipelines in the United States alone. Many of these pipelines run directly beneath the most densely populated places in the nation. Some are located in places where earthquakes, train derailments, plane crashes, construction accidents, landslides, floods, and other events will cause pipelines to rupture, resulting in explosions, fire, and hazardous materials spills that will endanger anyone nearby. Recent history is rife with examples of the insidious dangers created by buried pipelines. Since 1985, more than 325 people have died in pipeline ruptures and other related accidents.

Pipelines are such a danger that a special U.S. Federal Office of Pipeline Safety has been created to adopt rules to make these facilities safer. But the office has a huge challenge just trying to figure out where all the pipelines are. After six years of work, the Office of Pipeline Safety has yet to develop a national pipeline map, and some pipeline companies don't even seem to know where all their lines have been laid.

The rescue problems related to pipelines can be illustrated by six recent events. The first occurred in 1987 when a train carrying a soil-like material known as *trona* lost its brakes on the deadly steep Cajon Pass in California's San Bernardino County. The train roared out-of-control down the pass, leaving the tracks on a sweeping curve that sent the locomotive and cars crashing through a residential neighborhood. Nearly a dozen homes were wiped out by the train and buried in the powdery trona.

Twelve hours later, after many had written off any possibility of survivors, a USAR search dog called to the scene at the urging of Deputy Chief Mark Ghilarducci of the California OES found a man buried alive beneath tons of trona in a collapsed home. Two days later, as heavy equipment worked to remove the wreckage, a natural gas pipeline buried adjacent to the train tracks was punctured by a crane. The ensuing explosion killed two bystanders and burned several homes to the ground. If the pipeline had been ruptured during the initial train crash, the first day's rescue problems would have been spectacularly difficult, and several of the victims rescued that day might have perished.

When the Northridge earthquake struck Los Angeles on January 17, 1994, one of the more notable effects was the rupture and ignition of a major natural gas pipeline beneath Balboa Boulevard, within two miles of the quake's epicenter. Several homes burned to the ground, and several people were badly burned, including the man whose pickup truck ignited the inferno as he drove to check on the welfare of his mother. Not only did the blowtorch of flame threaten to cause a conflagration, drawing critical fire and rescue resources away from other life-saving operations, but it directly impacted some SAR operations in quake-damaged structures nearby.

In Chicago in 1998, a gas main rupture sent a towering pillar of fire ten stories into the air adjacent to a high-rise building housing elderly people. Flames directly impinged on the structure, threatening to involve the entire building. The situation exposed firefighters to a number of challenging fire and rescue problems that day. Only fast work by the Chicago Fire Department saved the lives of threatened residents. This kind of rupture-and-ignition sequence is a huge threat in modern cities in quake-prone regions. A fire of this magnitude next to an occupied building (especially one that's partially collapsed) would directly impact SAR operations and might force the IC to choose between fighting the fire and rescuing people trapped in the rubble. His decision would be made more difficult if local water mains were also ruptured, robbing his troops of adequate firefighting water supplies.

In 1999, an underground pipeline leaked gasoline into a creek near a subdivision in Bellingham, Washington. Children playing nearby went to investigate the smell of gasoline. Three were burned alive when the gasoline fumes found a source of ignition, causing a fireball that torched the entire creek.

On August 19, 2000, two families camping next to New Mexico's Pecos River were burned alive when an underground 31-in. (diameter) natural gas pipeline ruptured and sent a blowtorch of flame directly into the campground. The gas company noted a drop in pressure several minutes before a company employee hunting in the wilderness area saw the fireball erupt in the distance.

The previous incidents illustrate how operating pipelines can cause fires and explosions when they rupture, creating or complicating SAR operations. Planning for disasters of this type should begin, when possible, during the pipeline construction phase. Understanding where the pipe will be laid helps fire and rescue agencies determine the populations that will be at risk, and the local conditions (e.g. earthquake faults, landslide areas, construction zones, etc.) that lend themselves to pipeline rupture.

Terrorism-related disasters

One of the most pressing issues of the new millennium is assessing and preparing for the ever-growing potential for urban search and technical rescue disasters resulting from acts of terrorism. This century and the 20th century witnessed unprecedented acts of terrorism in the United States, including the September 11, 2001 attacks in New York, Washington (Fig. 2–13), and Pennsylvania. In 1993, the World Trade Center was bombed. In 1995, the Alfred P. Murrah Federal Building in Oklahoma City was bombed. Several terrorism campaigns were conducted by lone serial bombers and an Arab newspaper building in California was bombed. White supremacist organizations perpetrated a bank robbery and bombing. Various radical groups of the 1960s and 1970s committed terrorist acts. The L.A. Times building was bombed in the 1920s, resulting in a Los Angeles firefighter becoming trapped by a fallen beam while fighting the fire that raged afterward.

To escape, the man's leg was amputated so that he could be removed. A scattering of other terrorist acts resulted in building collapses and other urban search and technical rescue emergencies.

Fig. 2–13 Assessing the Collapse to Develop the Operational Plan at the Pentagon on 9-11.

Terrorist bombings also caused many deaths and resulted in lengthy urban search and technical rescue operations in places such as Buenos Aires (collapsing an Argentine Jewish Center), South Africa, India, Britain, France, Lebanon, Israel, Sri Lanka, and other nations.

The late 20th century also brought a host of terrorist plots in the United States that were thwarted by law enforcement efforts or failed to materialize for a host of other reasons. Included were the plans of a white supremacist group to bomb several federal buildings in Arizona, Nevada, and California simultaneously (broken up by the FBI), plans of Middle East terrorist groups to bomb the Holland Tunnel and other New York City sites, and others too numerous to list.

In January 2001, a lone American man drove a tractor-trailer across the lawn of the California state capital building at 70 miles per hour, barreling into the structure and causing a fire and explosions that killed him, caused the evacuation of the capitol, and caused millions of dollars in damage.

The collapse of the World Trade Center twin towers (Fig. 2–14) caused almost every imaginable collapse rescue problem, including the loss of hundreds of firefighters, police, and other first responders. Terrorism has emerged as the cause of some of the worst collapse disasters in the past 100 years.

Fig. 2–14 World Trade Center Collapse

Determining the level of threat from such events in any particular region requires an assessment of a wide array of variables, including the current and expected political climate, local demographics, recent history, the presence of groups likely to harbor or support terrorist activities, and the presence of targets that terrorists might find attractive or suitable.

Tsunamis

As mentioned earlier in this chapter, scientists are now warning of the possibility that a series of very large tsunamis could wipe out extensive areas of densely populated coastline just minutes after the next big earthquake in southern California. This is in addition to the long-recognized potential for tsunamis to strike the Pacific northwest coast, and the newfound hazard of huge tsunamis that can strike the Atlantic Coast of the United States within hours of catastrophic events like volcanoes and underwater landslides along the Atlantic Coast of Africa and southern Europe. Tsunamis can cause tremendous damage even to sparsely populated areas. When they strike coastal cities, the results can be devastating (Fig. 2–15).

Fig. 2–15 Results of a Tsunami (Courtesy Professor Costas Synolakis, USC)

Marine disasters

The worst ferryboat disaster occurred in 1987, killing 1,500 people off the coast of Manila. The disaster occurred when the ferry was rammed by a tanker ship. In May 1942, more than 2,700 soldiers were rescued from the aircraft carrier USS Lexington as it sank in a battle in the Coral Sea. In 1944, more than 2, 600 people were successfully rescued while the *Susan B. Anthony* foundered off the coast of Normandy, France.

Case study 5: disaster planning for El Niño-related disasters

 One of the most significant disaster cycles of the previous century was the El Niño event that developed in the late 1990s. Tens of thousands of people perished around the world in hurricanes, floods, mudslides, landslides, and other disasters related to the warming of Pacific waters that mark the El Niño phenomenon. As scientists warned of the approaching El Niño event, American public safety agencies were compelled to develop emergency plans for storms the likes of which hadn't been seen since 1982 and 1983, when hundreds died in floods and other weather-related events in the Americas.

Long history of floods results in swift-water innovations

In California, fire/rescue officials were quite familiar with the dangers and problems that accompany El Niño events because the state had been racked by floods, mudslides, land-slides, mud and debris flows, avalanches, coastal erosion, and other deadly effects of the El Niño that arrived in the years 1982-1984. During the winter El Niño storms, several LACoFD firefighters entered fast-moving water to rescue victims. One firefighter nearly drowned attempting to rescue a boy from a low head dam in San Gabriel Canyon, and another had a harrowing experience when he was lowered off the upstream side of a bridge to attempt a rescue in the City of Industry. This author was assigned to a coastal engine company and was nearly drowned and injured during a successful on-duty rescue to pull a man from the ocean during a major El Niño storm. The man was washed off a rock at the base of a 250-ft cliff and was being beaten on the rocks by 12- to 15-ft waves in an inaccessible area where no lifeguards were available. These and other rescues demonstrated the need to enhance the fire department's ability to safely and effectively manage water rescue incidents.

The use of helicopters to pluck victims (and, if necessary, rescuers) from fast-moving water becomes critical in some water rescue situations. In Figure 2–16, L.A. County firefighters and pilots assigned to USAR and Air Operations practice helo-swift-water rescue in preparation for flood rescue operations. A simulated victim, a rescue-trained USAR firefighter, has been captured by a rescuer who is preparing to clip himself and the victim into the rope dangling beneath the helicopter, while the helicopter matches their downstream speed.

Fig. 2–17 Swift-water Rescue in Los Angeles

Fig. 2–16 Practicing Helicopter Rescue Operations

After connecting himself and the victim to the rope via a carbiner, the rescuer signals his readiness to be extracted. The pilot, guided by constant instructions from the helicopter crewperson positioned at the door of the cabin, begins to ascend while maintaining a downstream speed matching the current. The victim and rescuer are lifted from the water with little resistance (Fig. 2–17). This is one of many new swift-water rescue evolutions developed by Los Angeles area firefighters in response to problems caused by El Niño storms and other flood rescue problems.

In 1983, LACoFD Fire Chief (retired) Clyde Bragdon, in an attempt to identify the level of hazard presented by water-related disasters, formed a departmental water rescue committee to study the scope of the problem and recommend solutions. After intensive research and development, the committee recommended a program of equipment distribution and training for high-risk areas of the county. The chief agreed, and the department began implementation of the water rescue program. The program included training and equipping dozens of engines, paramedic squads, helicopters, and truck companies. Prior to that time, there were few departments anywhere in California with formal swift-water rescue programs; swift-water rescue was not considered an official function of the fire service in that part of the nation. Prior to 1982 in L.A. County, formal training and response programs for swift-water rescue were practically unheard of among firefighters.

In the years 1992–1995, El Niño reappeared with even more devastating results. The storms were more drenching and violent, more homes were destroyed, and more people died in spite of the fact that local fire departments were better prepared. The flip side was that more people were successfully rescued, and firefighters and lifeguards were operating in these high-risk conditions with a higher degree of safety and better rapid intervention capabilities to back them up.

Floods of 1991/1992 prompt development of swift-water teams

In February 1992, southern California was hit by devastating floods that killed a dozen people in four counties. On February 9, the rain began falling and did not stop for seven days. As heavy rain from a Pacific storm continued to pelt southern California for a second day, indications of major rescue problems began to occur. Early in the morning of February 10, intense storm cells created deluges that overflowed creeks, threatening homes and stranding people in a wide area of the coastal foothills. Many of the initial problems occurred in areas traditionally prone to flooding. By dawn, however, flooding and rescues began occurring in places not accustomed to such trouble. This was just the first indication of the serious nature of things to come.

By mid-morning, 911 switchboards began lighting up. Many victims were becoming trapped by flooding creeks, and homes were beginning to wash away in some canyon areas. The LACoFD handled several rescues without mishap, and sand-bagging operations by the

department's fire camp system were in full swing in vulnerable areas of L.A. County. Other fire departments responded to swift-water rescues and successfully handled them. The situation seemed for the moment to be within the capabilities of the local response systems.

By late morning, however, conditions worsened dramatically. Intense storm cells were stalled over the Santa Monica Mountains and the San Fernando Valley. More than 6 in. of rain fell in less than three hours. Torrents of rain roared down normally bone-dry canyons toward the headwaters of the Los Angeles River. Visibility was cut to a few ft in some areas as clouds lowered almost to the ground and the downpour intensified. Water in some channels exceeded 30 miles per hour. The channels were reaching their capacity to carry water, and flood basins in the upper reaches of the canyons filled to their limit. Streams and flood channels suddenly began spilling over their banks in many areas. In the San Fernando Valley, home to many disasters over the years,[14] rescue and distress calls began pouring into the 911 system. People were being trapped by fast-moving floodwaters. A backlog of calls for fires and rescues quickly grew: Dispatchers began giving single engine companies as many as five simultaneous incidents; the captains were left to triage the most life-threatening cases and respond to them.

The Sepulveda Basin incident

At approximately 10:30 a.m., a completely unexpected event left no doubt that serious trouble was at hand. Northwestern tributaries of the Los Angeles River overflowed into the Sepulveda flood basin, which encompasses several hundred acres. The basin, normally dry year-round, is the site of a major park, a golf course, and other recreational facilities. These facilities include permanent buildings in areas never truly expected by most people to be under water. A major street, Burbank Boulevard, runs directly through the basin. In rare instances when water backs up enough to affect the road, gates are closed to block access and traffic is diverted onto other surface streets.

On the morning of February 10, Burbank Boulevard was heavy with traffic as cars slogged through the drenching rain. Without warning, a wall of water 3 ft high and hundreds of ft wide roared across the basin as the Los Angeles River finally overflowed. The flood smashed into a line of cars, sweeping some away. Within minutes, hundreds of cars were stranded and the water was rising over their roofs as the basin filled with water roaring over the banks of the Los Angeles River. Some motorists were able to race across the remainder of the boulevard to the safety of high ground. Others reached what they thought was sufficiently high ground but discovered to their terror that the water kept rising and eventually submerged the tops of power poles and streetlights. Many had to swim for their lives. Four dozen were trapped on top of their vehicles by the flood, which quickly rose over 25 ft in some places and created a huge, moving lake.

The Los Angeles Fire Department (LAFD) and Police Department (LAPD) began receiving frantic calls from citizens reporting a major flood in the Sepulveda Basin. Initially it was thought that the reports might be exaggerated because few people had ever seen this huge area go under water before, and it seemed incomprehensible that the entire basin could fill so quickly. However, first-arriving LAFD units were quickly convinced. They found a nightmarish situation unfolding as the water continued to rise, trapping additional motorists. The extent of the flooding was so great (and the weather so bad) that firefighters could not see from one side of the flood to the other. They could not even begin to estimate the total number of victims who might be trapped.

LAFD engine and truck companies braved the rising water to rescue victims from the tops of cars, trucks, trees, and other high points. The operation quickly became dangerous for firefighters. Several fire units became trapped as the water rose over their windows, forcing personnel to climb onto the roofs of apparatus with victims they had just rescued. Still others had to swim for their lives. With firefighters and victims trapped on islands of high ground far from shore and the water still rising, it became clear that helicopters and boats were required to conduct many of the rescues.

For several hours, LAFD helicopters conducted daring hoist rescue operations under weather conditions that were horrible for such duty. Visibility was severely impaired by the pounding rain and low clouds. Yet news helicopters crowded in on the rescue scene, allowing the nation to watch the situation live. In one memorable scene, a victim was hoisted 80 ft into the air and then, just as he was being pulled into the helicopter, he suddenly fell back into the flood. He disappeared from sight for several agonizing seconds but reappeared. He was removed from the water once again and survived with broken ribs and other injuries.

A group of lifeguards from the LACoFD's Lifeguard Division responded from Santa Monica Beach with IRBs normally used for rescues in the surf and open ocean. They assisted with the rescue of citizens and firefighters from high points not accessible to helicopters. The lifeguard response proved to be a lifesaving operation.

LAFD Assistant Chief Tony Innes, the IC, had his hands full. More than 500 acres of land was suddenly flooded with more than 20 ft of moving water; the water was pouring over the spillways of the Sepulveda Dam and the Los Angeles River was nearly overflowing. Severe weather conditions hampered operations. There was no way to determine how many victims might be trapped or missing, and many rescuers were trapped in life-threatening positions. Other fire and rescue incidents were draining resources in the region. After five hours of rescue and thorough searching, rescuers believed that all victims were accounted for. However, several hours later, a television news helicopter, preparing for a live afternoon shot, discovered yet another victim on floating debris hundreds of yards from shore. The news crew rescued this victim who was cold, tired, and wet, but alive.

Swift-water rescue teams established under emergency order

As rain continued to deluge the region, it became clear that the Sepulveda Basin incident was only the beginning of trouble. Even as victims were being plucked from the Sepulveda Basin, LACoFD Fire Chief Michael Freeman met with Operations Deputy Chief William Zeason to develop an emergency order establishing teams of swift-water-trained USAR personnel, including this author, who became the department's water rescue committee chairman and a swift-water rescue instructor. These firefighters and officers would be assigned to fire/rescue helicopters and ground units at the department's heliport and at strategically located fire stations across the county. The order was implemented that very day, and by 4:00 p.m., three swift-water teams were equipped, staffed, and ready to respond in helicopters and on the ground (Fig. 2–18).

Fig. 2–18 Practicing Helo-swiftwater Rescue Methods in Simulated Flood Channel

In the days to come, additional teams would be deployed to provide quicker response to outlying areas of the county. The helicopter swift-water teams augmented the existing two USAR/swift-water rescue-trained firefighters and pilot already assigned to the copters. Swift-water rescue ground units consisted of reserve brush fire patrol trucks that were pressed into service and equipped with swift-water rescue equipment from the department's training equipment cache. By 6:00 p.m., the teams were committed to life-threatening rescue situations as the deluge continued into the night.

Pasadena Glen incident

One of the first significant events of the evening came at approximately 8:30 p.m. when the LACoFD Fire Command and Control Facility (FCCF) began receiving 911 calls reporting a debris dam collapse and flood in an area of the San Gabriel Mountains known as Pasadena Glen.[15] The rain was coming down in such volume as to make it virtually impossible to see more that a few feet. Therefore, helicopter flight was ruled out and the swift-water teams responded in their ground units.

Members of the teams knew the Pasadena Glen well and could only guess that a debris basin had exceeded its capacity of rocks, mud, and water, and burst. In this scenario, it was easily conceivable that a wall of water 20 ft high might roll through the canyon, carrying away homes. Expecting the worst, they requested the response of an additional USAR truck company for rescue support.

As first-in, engine 66 approached the Glen on the single-lane road, they were met by panicked citizens self-evacuating the canyon and describing a huge flash flood. As engine 66 reached a point where the road drops down and crosses the normally placid stream, Captain John Nieto strained to see through the windshield, which was being pounded by rain. What he saw was a heart-stopping scene. Several cars swept past in the flood within 50 ft of the fire truck. Nieto could not see whether the cars were occupied, for they quickly washed over a 20-ft waterfall where the canyon drops away into a deep arroyo.

Across the flood, the lights of dozens of homes still shone, and Nieto could make out the forms of people moving about the shoreline. He assigned USAR ladder truck 82 and engine 11 to respond downstream of the glen to the Eaton Canyon debris basin to search for the cars (and possible occupants) that had been washed away. Meanwhile, engine 66's crew exited in the maelstrom and attempted to assess the situation.

It was clear that residents in the upper reaches of Pasadena Glen were stranded. Apparently, telephone communications had been severed. The only way to communicate was by yelling above the roar of the flood. This proved to be very difficult, if not impossible, until the flood later diminished. Nieto gathered from arm gestures and faint voice contact that some residents might be missing and others were moving to high ground in case larger walls of water swept the canyon.

The two newly formed swift-water teams, several engine companies, and the Battalion 4 chief arrived at the Glen. Nieto assigned the engines to work with the swift-water teams to conduct initial SAR operations in the Glen while truck 82 and engine 11 searched the debris basin. As the swift-water teams suited up, it was apparent that this operation would last hours and might be extremely hazardous. No one was exactly sure whether a debris basin had in fact failed or this was a typical flash flood for the area. It soon became apparent that a set of debris basins was still intact above the glen, but they were full and their ability to resist failure in the event of additional walls of water and debris was highly questionable.

Visibility was near-zero, and additional flash floods could easily take all rescuers by surprise with nowhere to run. Therefore, it was decided to make assignments carefully and maintain constant escape plans. A firefighter with a handi-talkie was posted upstream in a high ground position to give immediate warning of flash floods. The main emphasis was to conduct a rapid sweep of residences and look for victims who might be trapped anywhere in the glen. All residents would be directed to evacuate until the high danger period passed. Those isolated on the opposite shore would be directed to seek the highest ground until technical rope rescue systems could be established over the top of the flood for firefighter access.

All personnel would wear swift-water safety gear. A safety officer, search group leader, and rescue group leader were designated, and a communication plan was established. A liaison was also named to work with other agencies that Nieto had requested, including the L.A. County Sheriffs, Pasadena Fire Department, Sierra Madre SAR Team, and others. A command post was established at a local school.

Firefighters were organized into search squads and assigned to cover specific geographic areas. Each search squad included swift-water-trained personnel who would begin initial rescue efforts if endangered victims were found (Fig. 2–19). In several locations, technical rope systems were required to provide access to the steep, muddy banks of the flood. Truck 82 reported finding several vehicles that were not occupied. Due to the condition of the cars, there was no way to determine whether occupants had been swept out or the cars had simply been washed away from the driveways of homes upstream. Due to the volume of water and debris pouring into the basin and the weather conditions, full search operations would be impossible until morning.

Fig. 2–19 LACoFD Swift-water Rescue Team Training

After five hours, it was determined that most residents of Pasadena Glen were accounted for in safe locations. All endangered citizens had been rescued. Most of those who were able to evacuate the canyon did so without any prompting from fire officials. Swift-water rescue teams were made available to respond to other incidents, and it was not long before they were committed to rescues elsewhere.

Other significant incidents from El Niño

During the first days of the February 1992 El Niño storms, the newly formed LACoFD swift-water teams responded to many swift-water incidents, including the following:

- September 11, 1992—a man was washed down the Los Angeles River and rescued.

- Same day—several victims were washed into flood channels; some were able to escape on their own, and others required rescue.

- February 14, 1992—a car was swept away in Malibu Creek. Swift-water team members swam ropes across the creek (the only access available), then swam downstream to the vehicle. The car, now upside down under the fast-moving water, was searched and found to be empty. The transmission was found in drive and the ignition key was found in the on position.

- February 15—Two children decided to raft down a flood channel in La Canada. USAR truck 82, a helicopter swift-water team, a ground-based team, and other fire units chased the victims to a point where they could be rescued after emerging from an underground section of channel.

- February 15—A child was trapped in quicksand. He was rescued alive.

- Several citizens were rescued from the tops of cars and other entrapment situations.

Los Angeles River takes another life

Another significant event occurred on February 12, 1992 in the San Fernando Valley. This incident would result in a true *sea change* of philosophy in the southern California fire service, the public sector, and the political realm with respect to swift-water rescue and disaster preparedness.

The victim was a teenager riding his bicycle alongside a raging 20-ft deep concrete flood-control channel. The boy lost control of his bicycle, which slipped into the water. When he attempted to grab his bike, he slipped and fell into the churning brown current. The LAFD and LAPD sent a massive response to the incident. Firefighters and police officers spotted the victim at numerous points along the vertical-wall channel. They attempted to toss objects to him as he was swept past their positions, and he was able to grab several objects, but none could bring him to shore.

After several miles, the first channel converged with the 30-mile long Los Angeles River, which is 300 ft wide at that point and roughly parallels the Ventura freeway. The boy was forced to the center of the roiling river by the helical currents, which made shore-based attempts difficult. The water was approximately 20 ft deep, with 6-ft high standing waves. The river's walls were vertical in places and over 30 ft high. Under these conditions, rescue was nearly impossible without specialized equipment, training, and luck. Even where the walls transitioned from vertical to inclined, the tilt was nearly at a 40-degree angle, and the wet concrete made footing treacherous.

The news media took up positions on bridges, filming as fire units chased the boy more than 10 miles through the valley. Networks cut into regular programming to air the rescue efforts. Probably the most startling aspect of the entire episode was the multiple close-up shots of the terrified boy's face as the water carried him past rescuers who would not give up. For many, the sight was heart wrenching, one that would not allow the tragedy to be easily forgotten. For a time the boy clung to a floating log, finding temporary flotation. Meanwhile, engine and truck companies raced to downstream bridges and lowered ropes into the water. On several occasions, the boy grabbed the ropes. In one instance, he released his grip on the log to grab a line, thereby losing his only flotation. The water was simply too powerful, and he was immediately forced to let go.

Later, with a news crew filming, a police officer threw a coil of rope to the victim from shore. However, the officer's position upstream of a bridge made it impossible to pendulum the boy to shore, and the victim was immediately swept under the bridge and continued down the river. The effort was becoming desperate, and the boy was visibly weakened by the effects of his struggle with the roiling water and hypothermia. In one valiant attempt, an LAFD captain perched on the ledge of a bridge to attempt rescue. The water was literally coming up over the bridges; there was little room for a floating victim (or rescuer) to pass underneath without severe injury or death.

The next day, as the rain continued to pummel southern California, the Orange County Fire Department deployed two swift-water rescue teams at strategically located fire stations. The Orange County teams conducted several rescues. In one case, firefighters rescued several off-duty lifeguards who had attempted to boogie board a concrete, vertical-wall flood channel and became trapped in a strong hydraulic current.

One of the most dramatic examples of the success of the swift-water disaster preparations in L.A. County occurred in March 1992 when Firefighter Sean McAffee, assigned to Fire Station 69 in Malibu's rugged Topanga Canyon, rescued a trapped citizen by setting up a Tyrolean rope system over the raging flood-waters of Topanga Creek. McAffee, who had just completed the LACoFD's swift-water rescue training to become a member of the swift-water rescue team, was suspended from the Tyrolean system, maneuvered across the flood by firefighters using the rope systems, and plucked the victim from a flood-washed island in the middle of the river.

El Niño prompts development of multi-agency swift-water rescue system

As the floods of 1992 waned, the public continually focused attention on the need to provide firefighters with the best swift-water rescue equipment and training and to support fire department efforts to develop and maintain USAR and flood rescue systems. They also focused on the need to develop a multi-agency approach to swift-water rescue across L.A. County and southern California. As a result, the L.A. County Multi-Agency Swift-Water Rescue task force was formed to develop a coordinated system of planning, training, and response across the region.

The news media, which sometimes remained oblivious to the particulars of disaster response efforts, repeatedly noted the fact that fast, effective response to swift-water rescue incidents was not a result of happenstance. Effective response was the planned result of a 10-year program to improve swift-water response capabilities through training and equipping of hundreds of firefighters, by conducting research and development, and by continually striving to assure maximum effectiveness.

Some swift-water rescues require the use of sophisticated high-angle systems. In Fig. 2–21, firefighters assigned to LACoFD USAR units and swift-water rescue teams practice high-line methods to *pick off* victims trapped on cars, boulders, and other mid-stream objects.

Fig. 2–20 Swiftwater Rescue Team Training in Manmade River

The new emphasis on swift-water rescue resulted in development of new swift-water rescue teams by the LAFD and various other fire departments across L.A. County. The Multi-Agency Swift-water Rescue System was implemented immediately, and the result was a dramatic increase in saves as El Niño continued into 1993, 1994, and 1995.

During the 1993 storms alone, L.A. County firefighters, USAR companies, helicopters, and swift-water rescue teams rescued at least 111 citizens during more than 51 separate swift-water rescue operations. They prevented many millions of dollars of damage to homes and public property in Malibu and other parts of the county.

On one day, January 4, 1995, firefighters from the LACoFD and three other agencies rescued 200 people from fast-rising water when flood-control channels in the Carson area overflowed into surrounding neighborhoods, quickly swamping homes and automobiles. On January 10, 1995, LACoFD swift-water rescue teams rescued 35 people in the Malibu area. At least 12 people were rescued by fire department helicopters using one-skid landings, helicopter hoist cables, and new swift-water rescue/rope short-haul helicopter techniques. During the same period, 24 additional people were rescued by LACoFD personnel in other swift-water rescue incidents around L.A. County. These are examples of good disaster planning followed by successful implantation in times of disaster.

What would happen if a cruise ship caught fire or sank in waters off the U.S. coast? What would happen if a commercial airliner went down in your coastal waters with dozens or hundreds of survivors and deceased on the surface? Could your agency manage the consequences of such an event?

beneath the surface and are lost forever. In cold water, with injuries and no personal floatation devices, most victims have a narrow window of survivability. Many victims of marine disasters end up in the water in exactly that state. The most effective air/sea disaster plans include immediate response of available airborne and seaborne rescue resources. In Fig. 2–22, USAR firefighters practice their role as airborne rescue swimmers for marine disaster operations. In some cases, due to limited space in helicopters, the use of a short-haul (as depicted) is more appropriate for transporting multiple victims from the water to waiting watercraft that can accommodate them (as opposed to hoisting victims into the helicopter). This is a judgment call to be made by the rescue crews as they assess the disaster site, but it emphasizes the need for flexibility and the ability to adapt to rapidly changing conditions.

Fig. 2–21 Practicing High-line Methods

The most important issue in marine disasters is getting boat-based and aerial rescue resources to the scene as quickly as possible to provide flotation and rescue to victims before they slip

Fig. 2–22 Short-haul Rescue

Shipping disasters

In 1917, more than 1,600 people died from a blast in the hold of the French freighter *Mont Blanc*. The ship was packed with more than 5,000 tons of explosives when it collided with another ship in Halifax Harbor, Nova Scotia, Canada. The blast was felt more than 50 miles away. Then there was the Texas City (Texas) shipping explosion, which killed practically the entire fire department, leveled several square miles of urban area, and killed many residents. Are your port firefighters prepared for the consequences of such an event? What will be the SAR needs, and how will you manage them?

Requesting and Using FEMA USAR Task Forces

As of this writing, 28 FEMA USAR task forces are strategically located across the United States providing a timely and effective SAR capability when major disasters leave people trapped or missing. Each USAR task force consists of 68 specially trained firefighters, search dogs, paramedics, physicians, heavy equipment operators, hazardous materials specialists, communications specialists, and structural engineers. FEMA requires that each task force have three trained persons available to fill each of the 68 positions, for a total of 204 people.

The USAR task forces are in a constant state of readiness for missions anywhere in the United States,[16] complete with up to 42 tons of pre-packaged search, rescue, medical, and logistical equipment. In some cases, the FEMA USAR task forces are flown to disasters on military cargo planes to expedite their response. This was the case during the Northridge earthquake in which 13 FEMA USAR task forces were activated only moments after the quake struck. After the Oklahoma City bombing, 11 FEMA USAR task forces, including four from California, worked around the clock for 16 days to locate and remove all of the victims from the Alfred P. Murrah Building. At the 9-11 Pentagon collapse, up to 5 FEMA USAR task forces operated for 12 days; in New York, 26 of the nation's 28 FEMA USAR task forces operated at the collapse of the World Trade Center.

The USAR task force system was based, in part, on the pioneering work done by the OFDA, part of the USAID, within the U.S. State Department, which first dispatched members of the Miami-Dade (Florida) Fire and Rescue Department to the Mexico City earthquake in 1985, and then members of the Fairfax County (Virginia) Fire and Rescue Department to the Soviet Armenia earthquake in 1989.

Based on a mandate from Congress to upgrade the nation's ability to manage disaster arising from earthquakes, hurricanes, and other natural and man-made events, the FEMA system of domestically deployable USAR task forces was developed through the cooperative efforts of FEMA employees, private contractors, and many professional firefighters who worked together (often volunteering hundreds of hours of their own time with no compensation) to develop the national program. In fact, beginning in 1989, FEMA managed to establish this massive operation from concept to practically finished product in just over two years, an amazing accomplishment for any large federal government project.

The FEMA USAR task forces are designed specifically for the type of rescue operations seen in the Northridge earthquake aftermath. The USAR task force program had its beginnings in 1989 after the federal government's emergency response efforts to hurricane Hugo, the Loma Prieta earthquake, and quakes in Armenia and Mexico City exposed deficiencies in the ability of the United States to conduct rapid, large-scale, highly technical SAR operations in collapsed buildings. With these lessons in mind, FEMA coordinated the efforts of many agencies and individuals to create the USAR task forces, which are essentially a group of multi-discipline, rapid response disaster teams located strategically across the nation.

Understanding the FEMA USAR system in disasters

One of the main problems noted in recent disaster responses has been a lack of appreciation on the part of local officials for the capabilities and uses of the FEMA USAR task forces.[17] Some local officials across the nation are not yet aware of the program. At times, this has resulted in conflicts over deployment and command of the task forces, delays in recognizing conditions that require USAR task forces, delays in requesting them, and delays in assigning them where they are most needed. Some of these problems were seen in FEMA USAR responses to hurricane Emily, hurricane Andrew, hurricane Iniki, various tornadoes, and the Northridge earthquake. Fortunately, many of the conflicts are being resolved as more people become aware of the vast rescue capabilities of the USAR task forces. State and federal officials are stream-lining the process of activating and mobilizing them.

FEMA USAR task force response to the Northridge Quake

The Northridge earthquake provides a good example of how local officials' familiarity with the FEMA USAR task force system helps save lives when disaster strikes. In the early morning hours following the Northridge quake, it was apparent that some on-scene emergency personnel were not exactly sure if the conditions they faced warranted the request for USAR task forces; others did not know how to access the task force system; still others did not even know what a USAR task force was. Generally, this type of confusion can be reduced in future disasters by educating first responders and decision-makers about USAR capabilities available through local, state, and federal governments.

Fortunately, decision-makers in Los Angeles and Washington had been integrally involved with the development of the USAR program and reacted quickly to news of a major earthquake in Los Angeles. Immediately upon notification of the quake, OES Director Dr. Richard Andrews called his deputy chief, Mark Ghilarducci, who coordinates the state's USAR program and represents California in the FEMA USAR task force program.

Dr. Andrews instructed Ghilarducci to activate the California USAR Task Forces available to respond (of eight). It was automatically assumed that two of the six task forces (LACoFD and LAFD) would be unable to immediately mobilize full task forces due to the local impact of the quake. The other six task forces were dispatched to rendezvous with OES Assistant Chief Mike Douglas at Los Alamitos Armed Forces Reserve Center. OES/FEMA USAR task forces from Sacramento, Oakland, and Menlo Park were flown to Los Alamitos by military transport planes. The Orange County, Riverside, and San Diego task forces responded by ground. California National Guard ground transportation stood by at Los Alamitos, ready to move the task forces directly to any assignment.

Ghilarducci immediately spoke with Bruce Baughman, director of the FEMA USAR task force project in Washington, D.C., to discuss activation of task forces from other states. Baughman and FEMA USAR Program Manager Kim Vasconez conferred with other key federal officials. Nine FEMA USAR task forces outside California were immediately placed on alert. At the request of California OES, FEMA sent an 11-member incident support team (IST) to provide support to the state USAR response.

As damage reports came in, and aftershocks threatened to cause further collapses, FEMA activated USAR task forces from Phoenix (AZ), Montgomery County (MD), Fairfax County (MD), Metro-Dade County (FL), and Puget Sound (WA) to respond to March Air Force Base in Riverside County, California. The Phoenix task force responded directly by ground; the other task forces were flown by the Department of Defense. Within hours, 13 of the 25 FEMA USAR task forces were mobilizing to California, an unprecedented response to a disaster in the United States.

Planning for disasters in your jurisdiction

When James Lee Witt arrived at the site of the Oklahoma City bombing to help oversee the assistance being rendered by FEMA, in what was (to that date) the deadliest terrorist act in U.S. history, he knew that the massive SAR operations at the bombed-out Alfred P. Murrah Building were in capable hands; he had practically staked his career on it. Since 1992, Witt made a top priority of improving FEMA's ability to provide timely SAR assistance when state and local governments found themselves overwhelmed by the magnitude or scope of disasters. Ironically, during his time at the helm of FEMA, the United States was struck by more frequent, as well as larger and deadlier, disasters than at any other time in the nation's history.

Witt's successor, Joe Albaugh, was confronted by equally challenging disasters, including floods, an earthquake in Seattle, and the most lethal terrorist attacks and the largest structural collapses in the history of the world, events that took the lives of key members of the New York City-based FEMA USAR task force. Under Albaugh's direction, the FEMA USAR task forces continued their tradition of effective emergency operations under extraordinary circumstances. Albaugh resigned from the director post in 2003 to pursue other goals and was replaced by James Pauleson, former fire chief of the Miami-Dade County Fire and Rescue Department.

The cornerstone of the FEMA USAR system is a network of 28 USAR task forces strategically located across the United States to provide rapid response anywhere in the country or its territories in times of disaster. California weighs in with eight USAR task forces, owing to the preponderance of major disasters that occur there each year. Virginia and Florida are home to two task forces, respectively.

Each USAR task force consists of 68 specially trained firefighters, officers, paramedics, physicians, search dog teams, structural engineers, heavy equipment operators, hazardous materials technicians, logistics specialists, communications specialists, WMD specialists, and others with special SAR skills. These positions are designed to ensure the ability of the USAR task forces to conduct highly advanced SAR operations in cases of building collapses, explosions, tunnel collapses, hurricanes, tornadoes, major flooding, avalanches, mudslides, debris flows, plane crashes, and virtually every other conceivable type of disaster.

The California-based USAR task forces are capable of being broken down into several modules, including one or more 15-person swift-water rescue task forces, to conduct all manner of SAR operations during major floods. The swift-water rescue task forces are equipped with motorized IRBs, rope-throwing guns, equipment to conduct helicopter rescues, and an assortment of other rescue tools.

All FEMA USAR task forces are being trained and equipped to deal with the effects of nuclear/biological/chemical terrorism. The FEMA USAR task forces are arguably the most advanced disaster rescue teams in the world today, part of a national disaster management system that has proven itself to be highly effective in managing some of the most difficult SAR operations in recent history.

Each task force deploys with a cache of 50,000 lbs of equipment. The equipment is maintained in a constant state of readiness for immediate deployment. The equipment for each task force includes everything necessary to operate for at least 10 days, including food, water, shelter, and medical supplies, without any resupply for at least 3 days. This reduces the level of support that's necessary from local sources.

The equipment cache includes a wide array of specialized search tools, rescue devices, medical equipment, and support supplies. The equipment caches are packed in special waterproof containers, arranged on military-grade pallets, loaded on 40-ft trailers, and stored at the home base for each task force. Each task force is located near a military base, from which deployment can be expedited by flying in military cargo planes. In fact, FEMA requires that task forces be ready to be in the air within six hours of activation to a disaster. This helps ensure the most timely response possible for the nation's most advanced disaster rescue teams. In cases where distances are shorter and air transportation is not necessary, USAR task forces can generally be on the road within a few hours of notification.

The USAR task forces are fully staffed and partially funded by the local sponsoring agencies and states and are therefore considered to be local resources when disasters strike close to their home bases. In California, each of the eight task forces is strategically located in an active earthquake zone. When the Northridge quake struck, both the county and city of LAFDs were able to immediately activate their task forces and be assigned to the most urgent rescue operations.

The FEMA USAR task forces are also considered resources of their home states, available to respond to any state disaster at the discretion of the home state's OES. The California OES has utilized the task forces extensively in recent years. In June 1996, following a major rock slide that killed one hiker and left others missing, California OES activated the Sacramento USAR task force to assist the U.S. Forest Service with SAR operations. Most recently, California OES activated all eight of the swift-water rescue task forces to assist with dozens of major rescue operations stretching from Yosemite west to Monterey and north nearly to the Oregon border, areas that were flooded by torrential rains in December and January.

Similarly, the Florida OES has used its two USAR task forces to assist with SAR during several hurricanes and in the aftermath of the Valu Jet crash in the Everglades.

Virginia, California, and Florida have also been actively involved with providing international disaster assistance. Firefighters who would later become members of USAR task forces in California and Florida were sent to assist with SAR following the 1985 Mexico City earthquake, which killed an estimated 10,000 people. Members of the Virginia and Florida task forces were dispatched to deadly earthquakes in the Philippines and in Soviet Armenia.

Today, the USAID and OFDA are responsible for deploying up to three USAR task force under contract for international disaster response. Each of the USAID/OFDA international USAR task forces is sponsored by agencies that also deploy FEMA USAR task forces. Essentially, the same firefighters and rescuers that make up the three international USAR task forces also deploy as FEMA USAR task force members to domestic disasters.

For local officials who may be faced with disaster requiring the deployment of one or more FEMA USAR task forces, there are some key issues to consider. Most of these factors should be addressed in the planning stages to reduce confusion and delays when an actual disaster strikes.

1. By its very nature, the process of mobilizing a USAR task force may take several hours. All FEMA USAR task forces have a maximum 6-hr window to arrive at their point of departure for airlift. The most common method of transporting USAR task forces is via military aircraft, and each task force is assigned a specific military base as a point of departure.

2. Many task forces have the ability to respond directly to the disaster by ground within a certain radius. This may actually reduce the response time over an airlift and allows the teams to remain mobile, maintain equipment security, maintain communications, and does not burden the local

jurisdiction with transportation needs for the incoming task force. On the orders of Fire Chief P. Michael Freeman, the LACoFD USAR Committee developed a plan to mobilize its USAR task force in just 2 hrs for ground responses. The main purpose was to ensure quick response to quakes in L.A. County; any response within a 300- to 500-mile radius would also elicit a 2-hr mobilization. Ironically, the 2-hr mobilization plan was first used for the Northridge earthquake.

3. Even with improvements in quick response capabilities, the locality experiencing the disaster must anticipate that FEMA USAR task forces will generally take a *minimum* of 4 to 10 hrs to reach the point of arrival. Task forces from further away may take more than 15 hrs to arrive. It takes time to assemble and move such a vast amount of equipment (over 20 tons), personnel (56 persons, including structural engineers and physicians), search dog teams (4), and medical supplies (20 large boxes). Additionally, many medications and controlled substances have brief shelf lives and require special arrangements to procure and store.

4. The logistics of assembling and maintaining a FEMA USAR equipment cache in ready condition are enormous. Much of the early work on cache development was completed by members of the Dade County Fire/Rescue Department and the Fairfax County Fire Department. The LACoFD task force equipment cache has evolved through the experiences of Dade and Fairfax and its own experiences with multiple mobilizations and alerts. The entire cache is loaded into sturdy, waterproof boxes. Each box carries specific items designated for use by one of the major disciplines of the task force: technical search, rescue, medical, technical team, logistics. These boxes are then loaded onto pallets, which are loaded onto 40-ft trailers. The entire load is strapped down and fitted with covers.

This packaging system allows rapid loading onto military pallets for loading into military transport planes. It also enables rapid off-loading at the scene of an incident. The entire equipment cache is computer-inventoried and marked to ensure rapid access to any needed item. The system also enables tracking of any item throughout the mission.

5. To minimize wasted time, it is incumbent upon the local incident command to rapidly identify collapse sites that will require USAR task force operations to save lives. The sooner these task forces are requested and assigned to an operational area, the sooner rescue work can begin. FEMA may elect to stage USAR task forces in federal mobilization centers before local requests are received; however, the affected state must request federal assistance before the task forces can move to actual collapse sites and begin work. One difference is found in California, where OES can dispatch USAR task forces directly to incidents through the state mutual aid system. Because of its involvement as a participant in the FEMA USAR program, LACoFD command anticipated this and deemed such site surveys to be a major priority after the Northridge earthquake.

Conclusion

Sometimes the most remote-sounding disasters are the most devastating, not only due to the level of physical damage and life loss they cause, but because they often catch us unprepared. What are the chances, we might ask, of a major American metropolitan area being submerged beneath the ocean during an earthquake? Or buried by lava? Or wiped out by volcano-spawned mud and debris flows? Or struck by a tsunami? Or buried beneath a landslide?

For that matter, what were the chances of a terrorist attack involving suicide teams that hijacked four commercial airliners and crashed them into buildings, causing the collapse of twin high-rise buildings in New York, the partial collapse of the Pentagon, and an airliner crash in Pennsylvania? These crashes resulted in killing thousands of innocent occupants and hundreds of firefighters, police, and EMS workers.

What would be the appropriate response to pre-September 11 suggestions that such outlandish events might occur? On September 10, 2001, the very suggestion of such an attack may have seemed so far-fetched that many people—including many fire/rescue officials and emergency planners, not to mentioned terrorism *experts*—would have dismissed it out of hand.

Surely, we might think, there must be more pressing issues of the day than making elaborate preparations for long-shot disasters that will probably never happen. That attitude is exactly why some fire/rescue professionals and others who should know better are caught like a deer in headlights when the worst-case scenario happens. The problem—and the responsibility—for fire and rescue agencies is to go far beyond the *ostrich mentality* and actively seek out signs of impending disaster, and then develop constructive response strategies based on the best information and experience available.

Endnotes

1   For firefighters and rescuers who because of local conditions and/or assignment to specialized teams, respond to disasters on a regular basis, the *experience quotient* for rescue operations can be quite significant. Some firefighters and rescuers are so accustomed to conducting rescue operations under disaster conditions that functioning in the disaster environment and dealing with massive urban search and technical rescue problems becomes second nature. These personnel have a distinct advantage when faced with disastrous conditions because they have become conditioned to revert to the disaster mode of operation when faced with large-scale emergencies or disasters, and they may be more impervious to the *shock-and-awe* stage of most disasters.

2   Local terrain features such as mountains, rivers, canyons, waterfalls, tunnels, cliffs, coastal bluffs, and flood plains are good indicators of conditions that can lead to natural disasters.

3   Man-made features such as dams, subways, high-rise buildings, industrial complexes, refineries, bridges, and highways are indicators of potential man-made disasters.

4   A major hurricane is defined as a category 3 storm with winds higher than 110 mph.

5   ICS is also used by many fire/rescue and police agencies nationwide to provide organized management of all types of everyday emergencies involving multiple units.

6   The SEMS law was created in California in 1993 as a result of the East Bay Hills wildfire that killed more than 25 people (including two firefighters). The law was intended to ensure a highly organized and timely statewide response to all disasters.

7   Scientists also discovered two very dangerous earthquake faults running along the bottom of Lake Tahoe, whose depths have only recently been fathomed by unmanned submersibles and other specialized techniques. Some seismologists and tsunami experts are convinced that rupture of either fault may produce tsunami-like waves that can exceed 30 ft in height and ricochet back and forth across the lake like water in a

bathtub for many hours after the earthquake. Such an event would be devastating to the shoreline towns that line Lake Tahoe and could produce hundreds of deaths.

8   Gore, Rick: "Wrath of Gods, A History Forged by Disaster," *National Geographic*, July 2000.

9   Ellis, Richard, *Imagining Atlantis*.

10 The North Anatoli fault separates Anatolia from the rest of Eurasia and has caused nearly 600 documented earthquakes—40 of them magnitude 7 or greater—since the time of Christ.

11 *Torrance Daily Breeze*, "Hospitals May Be Unsafe in Quakes," March 29, 2001.

12 Rasmussen, Cicilia, "An Avalanche of Water Left Death and Ruin In Its Wake," *Los Angeles Times*, February 16, 2003.

13 *Guinness World Records 2001*, Guinness World Records LTD, London UK.

14 The San Fernando Valley is no stranger to disaster. The valley has been the site of tremendous flooding during the last century. The 1971 Sylmar earthquake and the 1992 Northridge earthquake were centered there, killing more than 120 people and devastating large portions of the San Fernando Valley. Several major wildfires have begun in the Valley and burned all the way to the ocean. The 1993 Malibu firestorm, which killed four and burned hundreds of homes, was only the latest. Also in the valley was the 1993 Chatsworth fire, which overran and nearly killed an LAFD engine company.

15 Pasadena Glen is a steep, narrow canyon at the foot of an extremely steep, nearly continuous 5,000-ft rise of the San Gabriel Mountains between the highland towns of Altadena and Sierra Madre. There is only one method of ingress and egress for dozens of residents: a single-lane road that literally becomes a river during heavy rains, isolating many homes from help. The area has been the site of many disastrous fires and floods. Most recently, dozens of homes were lost there during the firestorms of 1993, and the Glen was further devastated by floods and debris flows in the 1993/1994 winter.

16 As of this writing, three of these USAR task forces (Fairfax County Fire and Rescue Department, Virginia; Miami-Dade Fire and Rescue Department, Florida, and the LACoFD) are internationally deployable under the auspices of the USAID, OFDA.

17 Eight of the twenty-five FEMA USAR task forces are located in California. Under the California Master Mutual Aid Plan, California OES has the authority to assign USAR task forces to disasters in California in the same manner that it helps coordinate state fire resources for wild fires and other major incidents.

3

Developing Multi-tiered Rescue Systems

With today's unprecedented emphasis on rescue, including traditional technical rescue and USAR, fire departments across the United States and an increasing number of other nations have established rescue companies, USAR units, USAR task forces, ground-based (or helicopter-based) swift-water rescue teams, and other specialized rescue resources. Countless fire department-based technical rescue teams have sprung up in urban and non-urban areas alike (Fig. 3–1).

Fig. 3–1 Urban SAR Unit Ready for Rescue and Fireground Duty in Los Angeles County

Although it may seem natural to assume that rescue was always an integral part of the fire service, and while the average citizen might expect that specialized fire department-based rescue units have always existed in one form or another, the reality is quite the opposite. For millennia, firefighters and other rescuers learned to make the best of whatever good luck came their way. If they lacked specific training, equipment, knowledge, and experience for certain emergency operations, they simply did the best they could under the circumstances and capitalized on any good fortune that presented itself. They handled rescue emergencies by improvising and making the best use of available resources, without the benefit of formal rescue training and without tools and equipment that's considered standard today.

Without formal systems of training, equipment, and specialized units in place to conduct high-risk rescue, firefighters and rescuers were sometimes doing pretty well if they finished the job without getting themselves and/or the victim killed. Some fire departments continue to use that Spartan approach today, even though the fire service is gravitating toward establishing more sophisticated rescue capabilities that take advantage of new techniques, training, equipment, and technology to improve the safety and effectiveness of difficult rescue operations.

Solid mechanical aptitude, common sense, and the ability to innovate were among the first formal firefighter entrance requirements developed by the London fire brigades in the 1700s. London firefighter candidates were required to have specific types of experience that would allow the fire chief to hand-select men from the construction trade, the commercial and military marine professions, and other disciplines that relied on these skills combined with physical prowess.

The working theory developed by the London fire brigade was that men who built buildings and sailed the seas were accustomed to using ladders, ropes, and other tools and techniques that were applicable to the fireground. These men, the brigade's chief reasoned, were also apt to be more disciplined and accustomed to working long hours under harsh conditions like wind, rain, sleet, and darkness. The theory held true, and the London fire brigades led the way in professionalizing fire departments worldwide, in part because of the practice of selecting firefighters from the ranks of professions whose intrinsic skills and traits were readily transferable to fire and rescue operations.

Before the first specialized rescue company was set up by the Fire Department of New York (FDNY) at the turn of the 20th century, rescue wasn't considered a discipline of its own. Rescue didn't command attention worthy of committing an entire company of firefighters. This was especially true during difficult economic times when fire departments struggled simply to staff the normal complement of engine companies and truck companies. Until the advent of the first FDNY rescue company, the concept of a fire department unit whose job was specifically devoted to handling rescue emergencies was foreign, including even those situations in which other firefighters were in need of rapid intervention (Fig. 3–2).

Fig. 3–2 Rescue 3, Part of the FDNY's Special Operations Command

Although there were variations of Alpine rescue teams scattered across Europe and the United States, rescue was often an ad-hoc affair conducted by volunteers thrown together at a moment's notice to rescue someone in need. Rescue remained that way in most places in North America until well into the 20th century. In some regions of the United States and Canada, rescue wasn't recognized as a formal discipline or a responsibility of fire departments until the last decade or two.

Before rescue became a recognized part of the fire service, firefighters confronted with complex and highly dangerous rescue emergencies had little choice but to rely on their fireground training, their cumulative experiences, and on their good old common sense (perhaps the most important ingredient of successful rescue operations). In short, firefighters were often ill prepared for complex rescue emergencies, especially those they had never seen before. Still, they managed to persevere through a combination of willpower, a sense of duty, dedication to human life, and inherent mechanical aptitude and problem-solving skills that many rescuers naturally brought to the plate.

Unfortunately, despite the best intentions on the part of firefighters and other rescuers, trapped victims often did not survive the ordeal of waiting for rescuers lacking formal rescue training and experience to figure out how they were going to conduct the rescue. Many trapped victims simply lacked the time to stay alive while rescuers attempted ad hoc, untested methods that they threw together in the heat of battle. Many victims died because firefighters and other rescuers simply had no idea how to go about rescuing them from complex entrapments. Even when the rescuers did have good ideas about how to extract trapped victims, they often lacked the equipment to make it happen while the victim was still alive.

The Rescue Revolution

The east coast and midwest regions of the United States led the charge for improved rescue through much of the 20th century by adopting the fire department rescue company concept and other advances that began the formalization and standardization of rescue. The rescue revolution began when the FDNY recognized the need to establish a unit with the tools and personnel to rescue firefighters who became trapped in the collapse of burning buildings, as well as other situations requiring the kinds of skills and tools that have become part and parcel of today's rescue companies. The success of the FDNY's rescue company concept convinced other large fire departments to employ the same strategy.

Urban fire department rescue companies began sprouting up along America's eastern seaboard, particularly in cities with dense housing, large industrial complexes, and other conditions that resulted in frequent collapse and heavy fire operations. The idea eventually caught on with fire departments in more rural areas that found it advantageous to have special units carrying equipment for road accident extrication and other common rescue emergencies. The rescue company trend gradually spread westward and southward. As the decades passed, it became evident that specialized fire department rescue units were an effective way of handling emergencies involving trapped people.

Rescues and rescue companies were eventually established in select U.S. west coast fire departments such as those in San Francisco, L.A., and San Diego. In Figure 3–3, Jim Paige, former battalion chief and firefighter assigned to the fifties-era Rescue 11 company of the LACoFD, poses between his old ride and a more modern version of the rescue unit. However, many other departments in the west shunned the concept as unnecessary, relying instead on truck companies to conduct the bulk of heavy rescue operations and taking pride in the ability of versatile engine companies to manage the wide range of emergencies to which they were dispatched, even without specialized training for complex rescues.

Fig. 3–3 Old and New Rescues

The popular theory in many parts of the west was that basic fireground skills augmented by training for typical extrication equipment such as hydraulic tools and rope would enable the firefighter to find a way to rescue most trapped victims. This approach worked most of the time because many firefighters by nature come equipped with a certain level of mechanical aptitude and problem-solving abilities (Fig. 3–4).

Fig. 3–4 Equipment Carried by 1950s-Era Rescue Unit in LACoFD

When firefighters found themselves trapped and in need of immediate assistance to save their own lives, their colleagues were often unable to provide a higher level of assistance because they lacked the training, equipment, and experience to rescue their own co-workers in a timely manner. Even worse, the supervisors of these firefighters were often unprepared to recognize and *prevent* such tragedies because they lacked the training to understand the inherent hazards of rescues in trench collapses, confined spaces, floods, collapsed buildings, and other technical rescue environments (Fig. 3–5).

Fig. 3–5 Modern USAR/Rescue Companies Carry More Equipment than Predecessors

True to form, firefighters assigned to engine and truck companies with little—if any—formal rescue training have managed to rescue people from all sorts of predicaments they have encountered. They have managed to pluck victims from flood-control channels wearing full turnouts and dig people out of collapsed trenches without emergency shoring. They have plucked people off cliffs without sophisticated rope rescue gear and with little practical experience in high-angle rescue. Firefighters also managed to extract people from machinery and used perseverance and plain guts to crawl into collapsed buildings to rescue victims after earthquakes. In the process, it has been fairly common for firefighters to break safety rules routinely recognized today. But the bottom line was that they managed to *get it done* under almost any circumstance.

Unfortunately, some firefighters have died when unstable trenches collapsed on them during rescue operations. Some have drowned while attempting swiftwater rescues wearing full bunker gear. Still others have drowned after capsizing their boats in lowhead dams because they didn't understand the power of the water or have fallen to their deaths attempting high-angle rescue because their ropes failed. Many more have been killed in secondary collapses of buildings.

And then there are the original victims of these incidents, the citizens who—after they became trapped—often did not receive the best care that might have been available. If the local fire department had established formal rescue units to plan, train, equip themselves, and gain experience in managing the full range of rescue emergencies, victims' needs would have been anticipated. The simple fact is fire departments that had not established rescue companies or some other formal system of rescue (including requesting mutual aid from agencies with rescue capabilities) did not provide the maximum level of service that was possible.

In summary, the formal recognition of rescue (including the most effective rescue of our own trapped and lost colleagues) as a fire service responsibility has resulted in more effective rescue operations, particularly in large cities with fully dedicated rescue companies. It has also vastly improved the effectiveness of fire-ground operations by developing more well-rounded firefighters whose problem-solving abilities have been tested and strengthened, and by providing them with more equipment to perform special tasks. Fortunately, in the latter part of the century, the west coast fire service seemed to follow the lead of the east by discovering rescue as a distinct discipline deserving of dedicated staffing and specialized training

Modern Rescue Systems

Today we can say that many progressive agencies have developed multi-tiered rescue systems. There are a number of variations from agency to agency, based on a number of factors such as local terrain, construction, geography, weather, particular rescue hazards, fire/rescue agency budgets, and even local tradition. Some of these systems require that all firefighters be trained to at least the rescue first responder awareness level and that secondary responders (some of whom may be assigned to designated truck companies, squad companies, or other units) be trained to the rescue operational level. The systems may require that specialty unit personnel (rescue companies, USAR companies, etc.) be trained to the rescue technician or specialist level. Other fire/rescue agencies have developed mixes of these programs tailored to their own local hazards, resources, and needs. Other rescue systems are multi-agency or regional in nature. In any case, this trend toward more organized and expert fire department-based rescue capabilities shows no sign of slowing anytime soon.

As more fire departments establish formal rescue units and systems, nearby agencies become acquainted with the advantages of these enhanced capabilities and the trend for still more fire/rescue agencies to start formal rescue programs continues. As this phenomenon continues, firefighters and fire/rescue agencies will continue to perform highly technical rescues with increasing levels of timeliness, effectiveness, experience, and personnel safety and accountability.

Whereas in the past rescue was sometimes an ad-hoc affair conducted by well-meaning but inadequately trained and equipped personnel who still managed to get the job done by hook or crook, trapped victims now have a far better chance of survival because the fire service has formally recognized and adopted rescue as a main mission for which there should be formal standards and accepted methodology (Fig. 3–6). As a result of this new emphasis on rescue (as well as recent events like terrorist attacks that have highlighted the need for the best rescue services possible), some of the most sophisticated technology available is being applied to USAR operations. Elaborate rescue response systems are being established by fire and rescue agencies, and difficult rescue operations are increasingly being met with success.

Fig. 3–6 Higher Rates of Survival with Fire Service Rescue Mission

Recent improvements in timeliness and effectiveness of large-scale disaster SAR operations in the United States are even more dramatic, as evidenced by massive, timely, and expert emergency response to the World Trade Center collapse, the Pentagon attack, the Oklahoma City bombing, the Northridge earthquake, a barrage of hurricane and flood disasters, and other rescue-related disasters that have challenged the nation's fire and rescue services in recent years. Many state offices of emergency service have discovered the advantages of supporting local fire/rescue departments by coordinating statewide mutual aid systems that ensure timely reinforcement of local rescue resources.

Beginning in 1985, the USAID and the OFDA fielded the first U.S.-based, internationally-deployable USAR task force in partnership with the Miami-Dade (FL) Fire and Rescue Department. Several years later, OFDA added another internationally-deployable USAR task force based in the Fairfax County (VA) Fire and Rescue Department. This innovative system helped set the stage for the development of domestically-deployable USAR task forces under the aegis of FEMA when the need became evident the year hurricane Hugo struck the Atlantic coast and the Loma Prieta earthquake hammered the greater San Francisco and Oakland areas.

The USAID/OFDA teams from Miami-Dade and Fairfax County helped establish new standards that have since been adopted internationally. In 2003, USAID/OFDA added the LACoFD's USAR task force to the U.S.'s international disaster response system, establishing an international response capability from the west coast.

In Figure 3–7, Virginia Task Force I (VATF-I), one of two USAID/OFDA USAR task forces that also double as FEMA USAR task forces, ends their assignment at the Pentagon collapse with a somber flag ceremony. The experience of VATF-I, whose members respond to disasters around the world as representatives of the United States, is extremely valuable in collapse situations of this magnitude.

Fig. 3–7 Virginia USAR Task Force 1 Conducts Flag Ceremony

Today it's commonplace for the USAID/OFDA USAR task forces to join USAR teams from other nations at international disasters, operating in a system that facilitates a sort of unified command in which the teams conduct SAR in a coordinated fashion. Beyond the simple humanitarian considerations that fire/rescue personnel hold in the highest regard, there is yet another reason for maintaining substantial USAR capabilities. The USAID/OFDA USAR response system helps ensure a higher level of rescue capability in the event of terrorist attacks and other disasters that trap people in U.S. facilities abroad. This is an important consideration at a time when international terrorism is on the increase and with the ever-present potential for other disasters that affect U.S. facilities.

In the aftermath of hurricane Hugo and the Loma Prieta earthquake, the U.S. federal government, through FEMA, established its formidable nationwide system of 28 USAR task forces. These task forces respond to domestic disasters in a cooperative effort with the nation's fire service, supported by a disaster management infrastructure that has become a model of rescue efficiency and effectiveness for many nations around the world. The FEMA National USAR Response System is by any standard among the most effective programs ever launched by the U.S. government, and it continues to improve and adapt as the USAR task forces are called upon to help local and state fire/rescue agencies manage ever-larger, more complex, and more deadly disasters.

Today the FEMA USAR system ensures a rapid response of teams of well-equipped, highly experienced rescuers anywhere in the United States or its territories. Since 1991, the FEMA USAR task forces and USAR incident support teams that respond with them have been dispatched to so many disasters that they have built up a substantial base of experience and knowledge in the tactics and strategies required to effectively manage a wide range of man-made and natural disasters. As an added bonus to the sponsoring agencies, the experience gained by personnel assigned to the USAR task forces is ultimately transferred back to the home turf as these rescuers return to their own fire/rescue agencies from disasters around the nation.

As a result of the efforts of USAID/OFDA, FEMA, and a large host of fire/rescue authorities and practitioners around the nation, for the first time in American history a collective pool of knowledge and experience from disasters nationwide has been concentrated into teams of highly trained firefighters and rescuers who can be dispatched to any disaster in the United States within hours. The lessons learned by these personnel are being captured, documented, and taught in a systematic way that ensures they will be applied to future disasters by fire departments and rescue teams across the nation and in other countries.

Figure 3–8 shows members of VTF-1 at the Pentagon collapse following the 9-11 attacks. Five FEMA USAR task forces were assigned to the Pentagon incident, and 20 FEMA USAR task forces rotated through assignment to the World Trade Center collapse operations.

Fig. 3–8 VTF-1 Members at Pentagon Collapse

Today there are national standards (including National Fire Protection Association [NFPA] 1670 and others) that address minimum requirements for fire service rescue training, staffing, equipment, and systems. This level of standardization is unprecedented in the history of the U.S. International rescue standards are also being developed and applied. In total, these developments represent a remarkable revolution in rescue. They are demonstrations of the evolutionary progress of rescue, an evolution that is taking rescue to new levels that could scarcely be imagined just a decade ago.

The positive effects of the continuing emphasis on expanding and improving rescue capabilities are twofold. First, there is a widening understanding that rescue is—and will always be—a primary role of firefighters. Urban search and technical rescue is something that cannot be separated from the other roles that firefighters fill. And second, there is a realization that improved service delivery is a natural byproduct of improved rescue. In fact, we have found—almost by accident—that there is no way to provide effective rescue service without improving the way fire departments operate and without creating more well-rounded firefighters who are better prepared to manage difficult rescue scenes and fireground operations. Today's firefighter may wear many hats, including EMS, hazmat, fire prevention, and others, but the modern firefighter also wears the rescue hat, and he wears it all the time.

Who Is Responsible for Rescue?

Today in most parts of the United States, the local fire department is responsible for providing rescue service. Other nations are gradually following suit as more governments and taxpayers realize that the fire department's main mission of saving lives and property is more aligned with rescue than law enforcement's main role of enforcing laws and preventing or solving crimes. This is especially evident where tight budgets and economic downturns force decision-makers to prioritize law enforcement funding to deal with crime and its effects by putting more police on the street instead of duplicating the rescue services that are already provided by the fire department (Fig. 3–9).

Fig. 3–9 L.A. County USAR Task Force Firefighters Conducting Over-the-side Rescue Operations in the San Gabriel Mountains Overlooking Los Angeles

Despite all logic to the contrary, the question of who is responsible for rescue remains something of an open question in some areas of North America

and other parts of the world. In many places, there is no formally identified primary agency for rescue. In some communities, the fire department is the ad hoc rescue agency simply because there is no one else capable of doing it. In most cases, these local fire departments fill the role magnificently. However, a problem still remains. If no agency is officially recognized as the primary rescue service and no funding and resources are committed to rescue, there may be little impetus for local agencies to establish and nourish rescue systems that provide the best resources to help victims in a timely manner.

In some communities, fire departments have not yet been officially designated the primary agency responsible for rescue, particularly in urban and suburban areas where there are no mountain rescue teams or other formal rescue services. These areas may not have funds designated to train and equip local firefighters to perform rescue operations safely and effectively. Trapped people in these areas often rely on local first responders who may not be fully prepared to rescue them in a timely manner and without causing unnecessary complications.

Furthermore, without the benefit of specialized training and equipment or backup from rescue units that possess the proper training, equipment, and experience, first responders themselves may fall victim to secondary effects. Secondary collapse of structures and trenches, rope system failure in high-angle environments, atmospheric hazards in confined spaces and tunnels, etc. can complicate urban search and technical rescue emergencies.

In mountain communities where formal mountain rescue teams (usually volunteer) have been formed, the situation is often more hopeful. These teams are available for people who find themselves trapped on cliffs or in avalanches, rockslides, or floods. They are there to help those who have driven their automobile over the side of a mountain road or into a swollen creek or even a frozen pond.

At least in these communities there is an officially recognized and sanctioned system of rescue in place that has usually been time-tested. In many cases, mountain rescue teams work in conjunction with the local fire department, which may also be a volunteer agency, to effect the rescue of people trapped in mountainous terrain. Even *in town*, mountain rescue teams may be called upon to assist the local fire department with situations such as trench rescues, confined-space rescues, and high-angle rescues. In Figure 3–10, rescuers assist after a car went 400 ft over the side on a mountain road, trapping the driver inside his burning auto and sparking a brush fire. Firefighters were confronted by intense brush fire conditions and a man trapped in his car in a steep canyon requiring rope systems to reach him. This photo was taken after the fire had been controlled and while rescuers were searching the canyon to ensure that there were no additional victims ejected from the car when it tumbled into the canyon.

Fig. 3–10 Fire Department and Law Enforcement Rescue Units Operating Together under Unified Command

Over the years, however, it has become abundantly clear that urban and suburban fire departments must both be prepared to manage the kind of rescue emergencies that occur in urban environments. During the past four decades, the primary mission of the fire service (to protect life, the environment, and property) has expanded to include EMS, hazmat response, and other tasks for which there was no officially recognized primary agency. Along with this expansion of the fire service mission came a natural inclination of fire departments to conduct rescue with ever-more effective results. The expansion of the formal fire service role to include urban search and technical rescue represents a somewhat new paradigm in the West, but one with deep roots in the East.

Rescue in the Era of Terrorism

It can be said that the history of the fire service is divided into certain eras that reflect major philosophical, technical, and societal changes that have directly impacted the way in which fire departments do their work. Some of these changes include the evolution of military-based fire brigades, pumps, volunteer fire departments, fire hoses, aerial ladder devices, horse-drawn apparatus, paid fire departments, fire codes and ordinances, motorized apparatus, and self-contained breathing apparatus (SCBA) (Fig. 3–11).

Fig. 3–11 FEMA USAR Task Force Members Assessing Partial Pentagon Collapse after 9-11 Terrorist Attack

The 1970s saw the development of the ICS, the emergence of the nation's first comprehensive pre-hospital care systems, and, thus, the era of the paramedic. The 1980s have often been referred to as the era of hazardous materials response. The 1990s have been called the decade of rescue, or USAR, an era that has actually continued seamlessly into the new millennium. Considering the current state of affairs and what we can expect from the future evolution of terrorism, it might be said that we have now (unfortunately) transitioned to an era of terrorism and the use of WMD against both military and civilian populations by individuals and terrorist groups at home and abroad.

If the word *fortunate* can be used at all in relation to these developments, it is because the era has spawned an evolution of the fire service. In the past three decades, extremely elaborate systems have evolved to manage huge numbers of resources, provide prehospital care for mass casualties, mitigate the effects of hazmat emergencies, locate and rescue people trapped in collapsed buildings and other large-scale urban search and technical rescue situations, and do all four at the same time in the aftermath of a major terrorist attack.

Had it not been for the many disasters and mishaps and visionary people who spurred the fire service to develop and shape these elaborate systems, the fire service might have found itself woefully unprepared to address the consequences of terrorism when it struck in earnest in the early 1990s. As it turned out, the ICS and the SEMS that followed, the various emergency medical care systems (including mass casualty systems), the hazmat response systems, and the USAR systems, are well-suited to take on the challenge being thrust upon the fire service by the rapidly evolving nature of modern terrorism.

One tragic aspect of all this should not be lost upon us. Some of the true pioneers in the development of the nation's USAR systems—including those who for years presciently warned that the next era of the fire service would be characterized by domestic and international terrorism—lost their own lives trying to save others during the biggest and most lethal terrorist attack in history.

Cheating death

Prior to the westward spread of formalized USAR and rescue systems in the United States in the 1980s and 1990s (supported by national regulations like NFPA 1670), the survivability of victims and rescuers alike was often in question because firefighters and rescue teams in many western regions lacked the most basic safety measures to deal with the sometimes unforeseen hazards they faced.

When there were few if any formal rules for such things, many firefighters and would-be rescuers in the west regularly cheated death when they rescued victims using tactics that seemed consistent with common sense but which actually broke some of today's most basic rules of rescue. Some of us managed to avoid being caught unawares while entering non-shored, collapsed trenches to dig out construction workers. We've managed to avoid drowning after jumping into fast-moving rivers and flood-control channels wearing full turnouts (or with ropes tied around our bodies) to rescue victims. We've escaped with our lives after entering confined spaces to rescue downed workers without respiratory protection. We've avoided falling to our deaths during high-angle rescue attempts when we used one rope instead of two or when we neglected to use any form of fall protection measures. We've avoided being squashed by secondary collapses after entering earthquake-damaged buildings without even the most basic shoring. We've survived standing on the skids of hovering helicopters without any form of attachment, and we've been hoisted from helicopters without someone else double-checking our harnesses and rigging. We've committed all manner of other violations of basic rescue rules without being killed.

Too often, we in the fire and rescue services cheat death even when we've broken the most basic rules. After we do it a few times, we may begin to feel that we're a bit invincible, or we downplay the times when we simply got lucky and perhaps didn't even realize it was providence that prevented a catastrophe. Nevertheless, every so often the odds catch up to us.

In the years before formal rescue training, protocols, and units were established, many firefighters—including some of this author's own colleagues—were tragically killed in preventable mishaps when non-shored trench walls buried them while they tried to dig out some hapless construction worker. Some drowned as their turnout gear dragged them under fast-moving water trying to rescue a child. Some fell to their deaths when rope failed catastrophically while attempting to rescue people from burning buildings or from mountain cliffs. Others fell from helicopters while attempting to pluck unwitting victims from cliffs or floods. Some were crushed by secondary collapses while attempting to tunnel their way to victims after earthquakes. Still others succumbed to toxic gases or oxygen-deficient atmospheres while trying to rescue people trapped in confined spaces.

A prudent risk management approach to rescue

Even when we're doing everything according to the latest standards and even when it seems we're taking the most prudent actions, things can go badly wrong. Sometimes there's no one around to help when it matters most. Mishaps and near misses should serve as a warning of the serious nature of rescue. They should remind us of the importance of developing multi-tiered response systems that ensure the highest level of rescue training, equipment, procedures, and safety measures applied to rescue incidents. Unfortunately, we often treat near misses as events to keep quiet about, as if it's better not to acknowledge how close we came to

catastrophe. It's one reason that so many agencies have developed strong institutional amnesia about mishaps and near misses, and it's one reason that we continue to see the same pattern of mistakes, oversights, and mishaps.

The modern fire/rescue officer must take a different view of near misses. Instead of burying or ignoring near misses, they should use them as a sort of planning tool. Think of near misses as a guide for correcting hazardous flaws in the local system before a catastrophe forces the issue to the forefront in unpleasant ways that no one wants to experience. In the nation of Turkey there is an ancient tradition of hanging a symbolic ceramic ball with a three-colored eye in homes and other places where good luck is desired. The eye, it is said, represents evil. The idea is to look evil straight in the eye, because evil cannot survive direct exposure to scrutiny. In much the same way, it's necessary for fire/rescue personnel to look near misses and mishaps straight in the eye, to uncover problems that may come back to haunt us if we ignore them and sweep them beneath the proverbial carpet. One way to correct flaws in rescue capabilities and to improve the risk management situation related to rescue is to ensure that effective, multi-tier rescue systems are in place.

Advantages of a formal rescue system

Probably the most important feature of modern rescue in North America is the adoption of fire department-based, multi-tiered USAR and rescue systems, which are defined by the following features:

- These rescue systems are based (in part) on well-trained first responders who are prepared to begin and conduct rescue operations at a basic level, even before the arrival of rescue companies, USAR task forces, and other formalized rescue units (Fig. 3–12).

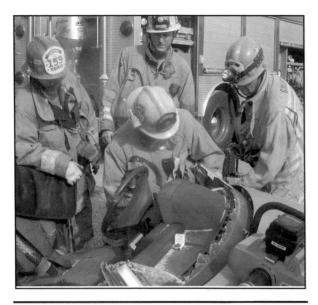

Fig. 3–12 USAR Firefighters Practicing Vehicle Extrication Techniques and Testing Prototype Battery-Powered Hydraulic Rescue Tools

Well-trained first responders are supported by more advanced tiers of secondary responders (e.g. rescue companies, USAR task forces, swift-water rescue teams, helicopter rescue teams, etc.) to ensure that higher levels of training, equipment, and experience are applied to complex or high-risk operations. Advanced rapid intervention capabilities are also made immediately available.

• Well-defined standards for rescue skills, equipment, tactics, and strategy have been adopted.

• There are measurable improvements in safety for victims and rescuers alike. There is compliance with the following standards and regulations:

  – NFPA 1670, *Standard on Technical Rescue*

  – NFPA 1006, *Standard on Professional Competencies for Technical Rescue*

  – NFPA 1983, *Standard on Life Safety Rope, Harnesses, and Hardware*

  – NFPA 1951, *Standard on Protective Ensemble for USAR Operations*

  – NFPA 1470, *Standard on Search and Rescue Training for Structural Collapse*

  – NFPA 1500, *Standard of Fire Department Occupational Safety and Health Programs*

  – NFPA 1561, *Standard on Fire Department Incident Management Systems*

  – OSHA 1910.134, *Respiratory Protection*

  – OSHA 1926.650, *Trench and Excavations*

  – OSHA 1910.146, *Permit Required Confined Spaces for General Industry*

The question is, how do we achieve these goals in an era of tightening budgets and competing interests?

Rescue system options

Fortunately, the options for effective management of urban search and technical rescue emergencies are nearly as limitless as the variety of fire/rescue systems that have evolved in modern times. Obviously there is no single *best* system, nor is there one system that addresses all possible rescue and disaster hazards and needs. But there are ways to adopt *best practices,* to mix and match smorgasbord-style to fit the capabilities to the local hazards, whatever they may be. Another successful approach is to ensure more effective coordination of *existing* emergency response systems. In other cases, development of entirely new rescue systems— including establishing new units—has proven effective. In yet other cases, an assessment and upgrade of existing rescue capabilities and systems can prove fruitful.

Rescue resource typing

Even with good planning and preparation, it's simply not feasible for every fire/rescue agency to have all the internal resources and specialized equipment and units necessary to properly and safely manage the full range of rescue emergencies that might occur. When disaster strikes, the resources of any single fire department may be overtaxed. The very definition of *disaster* indicates an event that may outstrip local resources. The concept of *mutual aid* in the fire/rescue services was developed to address both of these situations. Figure 3–13 shows an example of a mutual aid system available to respond anywhere needed to assist other agencies during major rescue operations.

Fig. 3–13 This USAR Company in an Urban Section of Los Angeles County is Part of the Region's Mutual Aid System

To have an organized system of mutual aid resources that can be called upon to perform designated operations or respond to certain hazards during large and/or complex rescue emergencies and disasters, it is first necessary to identify the *type* and *capabilities* of the resources available. This can pose a problem because of the simple fact that there is such a wide variety of emergency resources available in the modern world, each with its own set of characteristics, capabilities, and limitations. Resource typing is closely related to the ICS in one aspect: to request additional resources, especially those that require mutual aid response, an IC must be able to quantify the type and number of resources needed.

Examples

Looking at this from the fireground perspective, even the most common resource in the fire service world, the standard engine company, can come in many variations. The engine can be based on size, weight, pump capacity, water tank capacity, size and length of hose, 2-wheel drive vs. 4-wheel drive, ladders, rescue equipment, and staffing.

Suppose that an IC is at the scene of a wildland/urban interface fire. The IC needs 20 mutual aid engine companies to protect structures in a residential neighborhood and another 10 mutual aid engine companies to drive rugged dirt motorways into the mountains above the same neighborhood. The engine companies must stretch thousands of ft of hose through the hills in a flanking action. To accomplish these tasks, the IC must understand the capabilities and limitations of different types of engine companies or the IC risks losing the neighborhood because the wrong mutual aid resources may be sent and they may be incapable of performing the tasks.

The variations between different types of rescue resources are sometimes even more striking, and fewer fire/rescue officials are thoroughly familiar with the capabilities and limitations of each. If an IC at the scene of a structure collapse needs ten mutual aid rescue companies with certain capabilities and one or more FEMA USAR task forces to augment the rescue companies and other units to conduct round-the-clock operations, he needs to understand the characteristics of each of these resources or he risks losing the lives of trapped victims and rescuers alike.

To ensure that the right *type* of resources are being requested, dispatched, and used, there is a need to develop and use fairly precise definitions and standards when typing emergency resources for mutual aid, especially with regard to rescue. This process of characterization of resources is known as resource typing, and it is already an ongoing activity in many parts of the United States and other nations.

Once the resources of a city, county, region, state, or nation are typed, they can be inventoried with a reasonable degree of assurance that when an IC requests a certain type of resource, he will get it. Not only does this make life easier for the IC, but it's good for the people who are waiting to be rescued or whose property is threatened.

Typing procedures

Resource typing is conducted at various levels, including within municipal fire departments, within county or regional mutual aid systems, within states, and within federal fire/rescue agencies like the U.S. Forest Service. In places where mutual aid in some form is used repeatedly every day of the year, including states like California and New York where disasters have literally required the response and assignment of thousands of resources to manage the consequences of a single event, the typing of every fire/rescue resource available to the county, regional, and state mutual aid systems is important.

Looking back to the fireground example, we can examine the system of typing for engine companies. In some states, engine companies are typed by their pump capacity, water tank capacity, length and size of hose, and the number of personnel assigned to them. For example, a Type I engine company is essentially one that might be seen in any municipal fire department serving an urban or suburban area, with a 500-gallon water tank, etc., that can be assigned to conduct interior structural firefighting operations during disasters like urban conflagrations, or perform structural protection during urban/wildland interface fires. A Type II engine is one that carries less water, has less pumping capacity, and is built to deal with both urban and rural conditions. A Type III engine is one built for use on fire roads and off-road areas, with certain parameters for water tank size, pump capacity, etc.

Combining rescue resources into new types

In accordance with the ICS, similar resources can be grouped together in strike teams. For example, a Type I engine company strike team refers to five Type I engines that are assembled as one unit and supervised by a strike team leader who is typically a battalion chief with their own command vehicle. In California, for example, a Type I rescue company strike team consists of two Type I rescue companies and a battalion chief that responds as a single unit capable of performing the most advanced rescue operations.

Figure 3–14 is an example of a Type I USAR strike team, consisting of two like resources. In this case, two Type I USAR companies (one each from LAFD and LACoFD, respectively) join together as a task force for response to major USAR emergencies or disasters anywhere in California.

Fig. 3–14 Type I USAR Strike Team

Mixed resources can be grouped together as task forces to address particular problems requiring the use of mixed resources. Some task forces are permanent, with specific staffing and equipment requirements. Others can be assembled at the discretion of the IC to address specific problems or hazards.

Figure 3–15 shows a rescue task force that combines two different types of resources into one. In this case, a rescue-trained engine company (a rescue engine) is combined with an equally well-trained technical rescue company (e.g. a different type of resource than an engine company) to form a task force.

Fig. 3–15 Rescue Task Force Combines Two Different Types of Resources

For example, two Type 1 engine companies combined with a water tender can be dispatched as a fire attack task force. This type of resource is applicable in the aftermath of earthquakes, when water mains and other traditional sources of firefighting water may be disrupted. Similarly, two Type 1 rescue companies could be combined with a Type 1 engine company to provide water for fire protection and manpower during rescue operations and to supply water for concrete-cutting tools (to cool the blades), a Type 1 bulldozer (for emergency road clearing), and a commander, to respond to structure collapses in the aftermath of a damaging earthquake.

A more familiar example of a task force is the FEMA USAR task force. This is a team of 68 specially trained firefighters and rescuers who are assembled in times of disaster into five major components: command, search, rescue, medical, technical support. They are equipped to respond anywhere in the United States by air, ground, rail, or ship to complete a specific specialized task. That task is locating and rescuing deeply entombed victims of structure collapses and rescuing people in other forms of rescue-related disasters. Figure 3–16 shows FEMA USAR Maryland Task Force 1 operating at the Pentagon collapse.

Fig. 3–16 Maryland Task Force 1 Conducting SAR at the 9-11 Pentagon Collapse Disaster

An IC faced with the effects of a disastrous earthquake might be compelled to request mutual aid in the form of individual components, for example, two 50-ton cranes. The IC might also request larger increments, such as three Type 1 engine company strike teams, two Type 1 fire attack task forces, two rescue task forces, and two FEMA USAR task forces.

Typing of rescue resources: an example

For the purposes of this book, we'll examine the California USAR resource typing system as an example[1] (see appendix 1 for details). Certainly there are other states with equally representative systems for rescue operations, but the California system is most familiar to the author, and it's a system that's been tested by many rescue-related disasters and major rescue operations over the years.

The California fire service has developed multiple tiers of USAR capabilities and resources that are in addition to the state's eight FEMA USAR task forces. These levels have been designated basic, light, medium, and heavy, corresponding approximately to the type of structural collapse SAR and other technical rescue operations they are equipped, trained, and staffed to perform.

Because of the ever-present potential for catastrophic earthquakes in the state, California's multi-tiered rescue system was originally focused largely on response to earthquake disasters. But this system has proven that it's well-suited for a wide range of other disasters and major rescue emergencies, including dam failure, landslides, floods, mud and debris flows, avalanches, mudslides, tsunamis, explosions, underground emergencies, and terrorist attacks.

In California, the governor's OES is tasked with typing all fire/rescue resources and including them in the state's master mutual aid system, which divides the state into six geographical regions. To ensure consistency and accountability statewide, the inspection and typing of USAR resources is conducted by OES staff members of the special operations unit and the local OES fire and rescue branch assistant chief. They make on-site inspections of the resources, including a physical inventory of apparatus and all the equipment and personnel records to ascertain the appropriate category for mutual aid typing.

There are, of course, many other stellar rescue systems across North America that serve as models, locally, regionally, and statewide. Obviously it's impossible to mention all the most progressive rescue agencies here. But one that immediately comes to mind is the FDNY's system of five rescue companies (one rescue assigned to each borough), augmented by strategically located squad companies. It's no accident that the FDNY was the first American fire department whose leaders saw fit to establish a rescue company whose first role was to rescue firefighters from collapsed burning structures and other fireground mishaps, as well as locating and removing citizens from all manner of rescue predicaments. It's no less surprising that the basic concepts upon which the FDNY's rescue system is based have been mirrored (in smaller scale) elsewhere in the United States and around the world.

The rescue company concept is one that's proven its worth through the years, not only during technical rescue operations, but also on the fireground. Naturally, the concept of specially trained fire department rescue units has evolved according to the particular rescue hazards and needs of individual municipalities. Among agencies with progressive rescue company programs, one could list the fire departments of Philadelphia, Baltimore, Boston, Washington D.C., Fairfax County, Montgomery County, Virginia Beach, Dallas, Chicago, Houston, and many others.

In addition to rescue companies, there are other options for managing urban search and technical rescue emergencies. Following the lead of the FDNY, the Philadelphia Fire Department is—as of this writing—in the process of developing squad companies. The FDNY model of a squad company was pioneered in the late 1990s under the direction of Deputy Chief Ray Downey of the FDNY special operations command who understood the need to augment the department's rescue and hazmat capabilities to battle the threat of WMD terrorism and daily hazmat and rescue emergencies. In this model, squad company firefighters trained in both rescue and hazmat response were assigned to specially designed engine/pumper apparatus. They were then used as the second line of defense during both types of operations. The squad companies compliment the rescue companies by providing additional manpower and special skills because they are trained in disciplines like high-angle rescue, confined-space rescue, collapse SAR, etc.

In Figure 3–17, FDNY Squad Company 18 responds to a fire in Manhattan. In New York, squad companies combine select capabilities of FDNY rescue companies and hazmat companies with specially trained firefighters on a firefighting apparatus.

Fig. 3–17 FDNY Squad Company Responds to a Fire

The Miami-Dade Fire and Rescue Department is another example of a progressive rescue system. This is a department that—in recognition of the constant hazard of open waterways into which vehicles routinely plunge, as well as other submerged victim emergencies—has provided dive rescue capabilities on every single engine company, every truck company, and every rescue company. The department also has its own fleet of fire/rescue helicopters capable of dropping water on wildland fires, medevacing EMS patients from highway accidents, and transporting USAR personnel or divers to marine rescues and technical rescues on islands off the Florida coast. In addition to two heavy rescues, the Miami-Dade Fire and Rescue Department operates a USAR company that responds to all technical rescues and multi-alarm fires in the county.

Looking westward, the San Diego Fire Department has established a multi-tier USAR system consisting of first responders on all engine companies. The engine companies are supported by truck companies trained in basic USAR operations that are in turn backed up by Rescue 4, a fully-staffed rescue company.

The Orange County Fire Authority (OCFA) in California has adopted a system based on the concept of using existing truck companies as USAR units. The OCFA trains its truck companies to conduct a wide variety of

technical rescue and swift-water operations in support of the first responders. As of this writing the OCFA does not operate fully dedicated rescue or USAR companies as a third tier of response, but the department does employ two fire/rescue helicopters.

The Honolulu Fire Department's rescue stations respond to inland technical rescues and surf/ocean rescues around the entire island. The island has terrain ranging from crowded sun-drenched beaches to rainforests with magnificent cliffs rising many hundreds of feet above the earth, to high-rise buildings, to a major seaport and military facilities. The Honolulu rescue firefighters must be prepared to handle a variety of emergency situations, including dive rescues, mountain rescues, high-rise rescues, major collapses, and extrication operations. The same can be said of the Maui County Fire Department. These are examples of the diversity of rescue problems that face the American fire and rescue services.

Another example is found on the west coast, where the LACoFD, LAFD, Long Beach Fire Department, and a number of other fire/rescue agencies operate USAR units in a task force configuration. In this model, a USAR task force consists of one staffed USAR company that operates in concert with a specially trained USAR engine company, both of which are housed at the same fire station. The personnel assigned to both units are trained in rescue and respond together as a USAR task force to all technical rescues, to all multiple-alarm fires, and to USAR-related disasters.

For example, LACoFD USAR Task Force 134, assigned to a fire station in the northern half of L.A. County, consists of engine 134 and USAR 134 (a 54-ft heavy rescue apparatus on a fifth-wheel tractor-trailer configuration). Each unit is staffed with three personnel who combine to make a six-person unit at the scene of the emergency. Both USAR task forces meet California's standard for a Type I USAR company: six personnel trained and equipped to conduct the most complex USAR operations, including SAR in collapsed reinforced concrete buildings.

In Figure 3–18, LACoFD USAR Task Force 134 (USAR engine 134 and USAR Company 134) conduct high-angle rescue operations with USAR Task Force 103 (USAR engine 103 and USAR Company 103) in the San Gabriel Mountains above L.A. The firefighters at left are using a capstan system driven by a power take-off system in the rig. Electronic controls are used to haul a litter rescue team up the steep slope of the mountain. The mechanism uses a redundant safety system with double prussics and a prussic-minding pulley (here, tended by the firefighter on the right) to catch the load (rescuers) in case of a failure of the main line.

Fig. 3–18 Capstan System Delivers Rescue Team up a Mountain

USAR Task Force 134 is located in the desert north of the San Gabriel Mountains that divide the county practically down the middle. Task force 134 is the primary USAR unit for the northern half of the LACoFD's 2,300-square-mile jurisdiction. The task force is housed in a fire station in the L.A. Basin and its jurisdiction covers the entire southern half of the county, including the coastal zones and Catalina Island, 26 miles offshore.

It's a system that's proven itself over the years, and it works for a place with as many diverse rescue hazards as L.A. County. However, it may not work everywhere, and that is why it's critical to accurately evaluate the local hazards and needs before committing resources to a rescue program.

Development of USAR Systems and Other Rescue Resources

Readiness to manage the consequences of modern terrorism, disasters, and rescue-related emergencies is somewhat reliant on a comprehensive urban search and technical rescue program. This is not achieved with a Band-Aid approach. It is based on an ongoing commitment of significant resources to ensure the availability of proper equipment, training to address the local hazards identified in the daily rescue planning process, and continuing education to keep skill levels high. Readiness also involves adequate staffing to ensure timely and effective rescue operations and a plan to ensure that rescue resources are used to achieve the greatest advantage for the agency and the public. If these commitments are not part of the program, the results will be haphazard at best.

The most cost-effective fire department-based urban search and technical rescue system is one that fulfills multiple roles. Some of the most effective rescue units are those that respond to fireground as well as rescue emergencies. If a USAR program is to be established, ensure that it will enhance the agency's ability to fight fires and manage daily technical rescue incidents. If these tasks are closely related, you will get the best bang for the buck.

Consequence Management: Essential Backup to Damage Prevention

It is indeed true that society cannot be completely protected from the effects of disasters or daily rescue-related emergencies. When prevention fails to avert a rescue-related disaster or emergency, the next line of defense is good consequence management through timely, efficient, and expert response by the fire department and other rescue agencies.

Fortunately, our society has the ability to bolster its local, state, and national emergency response systems. This will ensure that when buildings fall down and massive fires start or many people are injured or trapped as a result of an earthquake, tornado, flood, or terrorist attack, there will be an organized, timely, and effective response.

Staffing Considerations for Rescue Units

It's reported that the famous explorer Earnest Shackleton interviewed nearly 5,000 men who applied for 50 posts on one of his most daring Antarctic expeditions. Shackleton sorted the applications into three piles: mad, hopeless, and possible. Shackleton handpicked each member for specific personal attributes, skills, and knowledge that he deemed necessary to deal with the harsh conditions of Antarctic exploration. Naturally, Shackleton favored men who were resourceful, dedicated, and perseverant, who had mechanical aptitude,

were seaworthy, willing to take great (but not foolish) risks, and capable of working as a team. It should come as no surprise to find that many of the traits Shackleton selected as important for his expeditions are also those that we attribute to good rescuers and rescue officers.

As mentioned at the beginning of this chapter, the first formal personnel qualifications for the London Fire Brigade included experience at sea, in construction, or other jobs that required men to work with ladders and ropes, to be comfortable at great heights, to be physically fit, and to be mechanically apt, resourceful, and capable of taking orders. Added value was attributed to characteristics like dependability, daring, the ability to remain mentally strong and balanced in the face of great difficulty, the ability to figure out difficult practical problems on the fly, and physical and mental endurance. Once again, these attributes are characteristic of those we find in effective rescuers and rescue officers.

The first fire chiefs in Europe and the United States had the relative luxury of being able to handpick their candidates. Even today, many fire departments (particularly some of those on the east coast) have provisions that allow rescue station officers to handpick members of rescue companies and other specialty units. It remains a tradition in some fire departments for rescue company officers to conduct a process of accepting applications, conducting career assessments, interviews, and even trial runs with the crew to determine which firefighters will be selected for duty on rescue companies, squads, and other specialized units. In some cases, the existing company members actually vote on new candidates.

It is, however, a fact that seniority rules in many fire departments. In some places, labor-management agreements specify seniority as the deciding factor.

Case study 1: LACoFD USAR system

 In the pantheon of fire/rescue operations, there are many agencies whose work is worthy of study because of their progressive approach both to daily rescue operations and rescue-related disasters. Time and space do not permit all of them to be mentioned here, but it's helpful to examine some representative agencies and their modern approach to rescue and to consider new paradigms that may at some point become the norm in some regions. Among the examples is the LACoFD, which established a multi-tiered USAR program in the mid 1980s and actively participates in regional, state, domestic, and international USAR response.[2]

With a population exceeding 10 million people living and working in a 4,000-square-mile area, L.A. County has a large population in one of the most diverse and disaster-prone regions on the planet. From west to east, the county stretches from Catalina Island (26 miles from shore) to the mainland with 90 miles of beaches and coastal cliffs. The county stretches inland over the Santa Monica Mountains, across the L.A. Basin, over several foothill ranges into the San Fernando, San Gabriel, and Santa Clarita Valleys. It extends over the 10,000 ft high San Gabriel Mountains, 650,000 acres of which is set aside as the Angeles National Forest that separates metropolitan L.A. from the north-south trending San Andreas Fault and finally spreads across hundreds of square miles of desert. Figure 3–19 shows L.A. County firefighters as a litter rescue team preparing to be lowered over the side of a 300-ft cliff along the rugged Pacific Coast of the Palos Verdes Peninsula.

Fig. 3–19 Mud and Debris Flows Present Multiple Hazards and Protracted Rescue Problems

Within this large and varied territory are hundreds of high-rise buildings, dozens of oil refineries, hundreds of chemical plants and factories, and thousands of industrial complexes of various types. There are also dozens of bridges over major rivers and harbors, subway tunnels beneath the city and train tunnels through entire mountains, many major tourist attractions, an international airport, three major domestic airports, and a dozen or so private airports. The county also includes two major shipping ports (including the largest in the United States), three marinas, nearly 100 miles of shoreline (whose beaches are outlined in places by 300-ft vertical cliffs and rocky outcroppings), and vast flood plains upon which have been built entire cities. There are countless politically-sensitive sites such as foreign consulates and government buildings, hundreds of gold mines drilled into the mountains and desert floor, innumerable flash flood zones, major community redevelopment and construction projects with deep excavations and ever-increasing building heights, and 500 miles of concrete-lined flood channels (some of which drop underground for long distances) whose water velocities exceed 30 miles per hour. The area contains a scattering of lakes and reservoirs with capacities up to 50,000 acre-feet, some of which were created by dams up to 400 ft high and are vulnerable to damage or failure from earthquakes, terrorism, and other causes. Some of the steepest mountains in the world are here, complete with ski resorts, high mountain rivers, deep canyons with waterfalls and high vertical cliffs, and cities built right up to the fringes of their steepest slopes.

Since the turn of the century, the greater L.A. region has experienced some of the most spectacular and deadly disasters in the United States. As of this writing, the L.A. area is identified with New York, Washington D.C., and San Francisco as among the most likely targets for future terrorist attacks (one of several emerging threats whose full impact has yet to be determined or experienced).

The infamous 1928 failure of the William Mullholland-designed St. Francis Dam in the Santa Clarita Valley area of northern L.A. County killed more than 500 people, making it the second-worst disaster in the history of California behind the 1906 San Francisco earthquake.[3] In recent years, floods, earthquakes, fires, tornadoes, mud and debris flows, landslides, and other natural disasters have severely tested the capabilities of the fire service there. The Long Beach earthquake in 1933 killed dozens and collapsed structures across parts of the L.A. Basin. In 1952, the Fort Tejon earthquake shook a four-county area with a magnitude of 7.7, making it California's strongest quake in the 20th century.

In Figure 3–20, L.A. County USAR firefighters prepare for tunnel rescue operations inside a cavernous reinforced concrete dam in a mountainous canyon above a desert community on the east side of the San Gabriel Mountains.

Fig. 3–20 L.A. County Firefighters Prepare for Tunnel Rescue

When the Baldwin Hills Dam collapsed, water washed away an entire L.A. neighborhood, killing citizens, and trapping firefighters attempting rescue operations. On February 9, 1971, the Sylmar earthquake struck the San Fernando Valley early in the morning, collapsing dozens of buildings across L.A. County, overturning a multi-story hospital, killing dozens of people, and threatening to kill thousands more when a large dam nearly collapsed. The same year saw a huge explosion in a tunnel under construction through the San Gabriel Mountains, killing nearly two dozen workers who were trapped inside, compelling L.A. firefighters to enter the tunnel in a high-risk bid to rescue the trapped men.

A series of deadly wildfires roared across L.A. County and the state of California in the early 1970s, prompting the development of the modern ICS. Then came a series of devastating El Niño-spawned storms that caused dozens more deaths across L.A. County, prompting the development of the LACoFD's pioneering swift-water rescue system that later evolved into the L.A. County multi-agency, swift-water rescue system, which became a model for other multi-tiered, multi-agency systems across the nation and around the world.

In 1987 the Whittier Earthquake collapsed buildings in eastern L.A. County, and LACoFD Fire Chief P. Michael Freeman responded by creating the department's USAR system. More years of disastrous flooding ensued, and the LACoFD's swift-water rescue and USAR systems proved their worth in the form of saved lives and prevention of rescuer casualties. The San Fernando Valley was struck once again on January 19, 1994 when the Northridge quake hit with a force registering 7.1 on the Richter scale, causing massive damage to a 400 ft high dam, collapsing several freeway overpasses that killed motorists and a motorcycle police officer, sparking several conflagrations, and collapsing many buildings with a death toll that went into the dozens (Fig. 3–21).

Fig. 3–21 Preparing for the Next Big One

L.A. County has a higher population than 42 American states, with an economy exceeding that of many nations. Practically every ethnic group, culture, religion, political persuasion, and language on the earth can be found in L.A. County. A number of groups with ties to domestic and international terrorism have established themselves in this region. The county is home to some of the world's great research and educational institutions and think tanks, including the Jet Propulsion Laboratories, Cal Tech University, the University of Southern California (USC), the University of California L.A. (UCLA), the Rand Corporation, and many others.

L.A. County is a place of varied microclimates and terrain, where one can find drastically different weather simply by driving over the next hill or into the next valley. From east to west, a person can drive across the flat desert of the Antelope Valley, up steep winding roads across the snow-covered and densely forested mountains of the Angeles National Forest, back down the mountains into large and densely populated valleys, across the cities of the L.A. Basin, and arrive at the ocean, within the space of two or three hours. More intrepid visitors can then hop on a ferry and travel 26 miles to Catalina Island. Indicative of the variety of terrain is the fact that some residents make a practice of surfing in the morning at the coast, then driving into the mountains to ski in the afternoon, and then back to the nightlife of the city.

In Figure 3–22, LACoFD engine companies are being transported aboard a U.S. Navy hovercraft to Santa Catalina Island. To address the unusual challenges inherent in providing fire/rescue service to an island 26 miles away from the mainland, the LACoFD and the U.S. Navy have developed an innovative partnership. When there are large emergencies on Catalina, Navy hovercraft from Camp Pendleton respond immediately to transport engines, trucks, USAR apparatus, wildland units, and other specialized equipment and personnel. This system is tested annually through exercises transporting LACoFD units to the island, and it has proven successful in controlling at least one major wildland fire in 2002. This unique system has become even more important with the emerging threat of terrorism on cruise ships that dock off the island and potential terrorist risks on the island itself, which is a major tourist attraction.

Fig. 3–23 LACoFD and the L.A. County Sheriff's Department Rescue of Scuba Diver from Submerged Mine

Fig. 3–22 LACoFD Engine Companies Going to Santa Catalina Island

Fig. 3–24 Rescue Diver in Submerged Mine (Courtesy of Mark Lonsdale)

Figure 3–23 shows L.A. County firefighters and sheriff's department SAR team members attempting the rescue of a scuba diver who went missing while exploring a submerged mine deep in the mountains above L.A. The eight-hour emergency operation required a combination of mine and tunnel rescue operations and cave dive rescue methods to recover the victim from a submerged chamber within the mountain. In Figure 3–24, a rescue diver makes his way through the underwater labyrinth of a submerged mine deep in the San Gabriel Mountains.

Earthquake Concerns Spawn USAR Program Development

To understand the diversity of rescue hazards found in L.A. County, one must recognize that the multiplicity of rescue emergencies roughly correlates to land use (e.g. cities, suburbs, townships, coastal towns, mountain enclaves, etc.) and to how people live, travel and work, which is somewhat dependent on the varied and ever-changing geological zones.

The changing terrain is the result of a colossal struggle between the forces of nature that create earthquakes and other features of the tortured terrain in greater L.A. As the Pacific tectonic plate grinds relentlessly past the North American tectonic plate, the earth's crust is fractured and folded into foothills and mountain ranges. The meeting point of these two huge plates is a 1,000-mile long rip in the earth's surface called the San Andreas Fault, one of the most deadly land features in the Americas. The San Andreas rises from Mexico's Sea of Cortez and runs inland of (and parallel to) the California coast, creating north-south trending mountain ranges in some places and transverse mountains in others. The San Andreas finally disappears into the Pacific Ocean north of San Francisco. The land between the two ends of the San Andreas is among the most earthquake-racked real estate on Earth.

As the Pacific and North American tectonic plates push against (and past) one another, the land fractures, creating hundreds of smaller faults in the earth's crust. The pressure created by this movement invariably causes periodic shifting of the faults, the result of which are earthquakes. Earthquakes occur in and around L.A. County virtually every day in varying degrees of intensity. Most are too small to be felt by people. Some, however, have caused massive devastation and loss of life.

The legendary 1906 San Francisco earthquake (which collapsed and burned most of the city and resulted in more than 500 deaths) and the 1932 Long Beach earthquake (which killed dozens, including students trapped in collapsing school buildings) taught many valuable lessons to architects, builders, and the fire service. Building codes were stiffened, building methods were improved, and the fire service began to prepare for large-scale disasters. Today, as a result of the better construction standards and reliance on stricter building codes, an earthquake of the 5 to 6 magnitude on the Richter scale may cause only light damage, or perhaps none at all. Whereas if a quake of the same magnitude were to strike cities in the midwest or the east coast, the results might be devastating.

The land beneath L.A. County has been broken, raised, lowered, and convoluted by the movement of the faults. Some of the faults are buried under densely populated valleys and alluvial plains and their locations may not be known until they rupture, as in the 1994 Northridge earthquake (Fig. 3–25) and the 1987 Whittier quake. Many of these faults are caused indirectly by pressures from the *dogleg* section of the San Andreas fault, which cuts through the eastern fringe of the San Gabriel Mountains. The dogleg causes the fault to lock in place for about 130 years at a time, exerting mountain-raising tectonic pressures on adjacent areas.

Fig. 3–25 Damage to Overpasses from Northridge Earthquake

Seismologists are convinced that the dogleg section of the San Andreas fault that divides L.A. County is overdue for rupture, and they warn southern Californians to prepare for a great quake in the coming years. When the fault finally breaks, it will rupture along perhaps hundreds of miles, with each side of the fault moving 20 to 30 ft relative to the other in an instant, causing an earthquake that may last up to four minutes. For anyone who's experienced even 15 seconds of violent ground shaking, the thought of such an event lasting four minutes is alarming. Scientists estimate that an 8.3 quake on the southern section of the San Andreas Fault may kill as many as 14,000 people in L.A. County alone. A 7.5 quake on the Newport-Inglewood fault may kill as many as 20,000 people.

Members of the LACoFD are keenly aware of these hazards. The 1971 6.5 Sylmar quake (more than four dozen people killed) and the 1987 6.1 Whittier quake (dozens of buildings destroyed) were centered in areas protected by the LACoFD. In 1971, the five-story Olive View Hospital fell on its side less than 1,000 ft away from LACoFD Station 46. Both the San Andreas and the Newport-Inglewood faults run directly through areas protected by the department. The disastrous Northridge and Long Beach quakes were also centered within the borders of L.A. County. With this history and the ever-present potential for future disasters looming, the LACoFD has been compelled to maintain constant readiness for such events.

Ironically, the earthquakes caused by these faults helped fuel the impetus for development of the discipline that's become known as USAR. Following the 1987 Whittier and 1989 Loma Prieta earthquakes, many California-based fire departments, along with agencies such as the California OES and California State Fire Marshal, developed *heavy rescue* systems that have evolved into the present-day USAR programs.

In response to local earthquake threats in addition to their other technical rescue hazards, the LACoFD developed a USAR company, and firefighters built an earthquake collapse training prop to simulate the collapse of a multi-story building (Figure 3–26).

Fig. 3–26 Earthquake Collapse Training Prop

Obviously, the potential for rescue is not limited to earthquakes. The LACoFD responds to every situation involving trapped people, whether in the city, the mountains, rivers and streams, or the ocean. The LACoFD provides firefighting, EMS, and technical rescue services in the Angeles National Forest, in the other mountain areas, in more than 60 cities served by the department, on Catalina Island, along the Pacific coast, and to all of the county's unincorporated areas. Their jurisdiction exceeds 2,300 square miles. The department's USAR, swift-water rescue, high-angle rescue, marine, and other technical rescue systems have been in place formally since 1983. This qualifies the LACoFD's rescue system as a sort of *fledgling* in comparison with the vaunted rescue companies of the FDNY and the rescue systems of many fire departments on the east coast and in the Midwest. Although the effectiveness of the LACoFD's USAR system has been proven during the successful management of a wide array of technical rescues and disasters, it continues to undergo vigorous reevaluation and revision based on lessons learned during both successful and problematic missions over the years.

Prior to the mid-1980s, without the benefit of formal rescue training, equipment, procedures, and without the personnel safety precautions that evolved in the early part of that decade, firefighters in L.A. County (like many other regions of the United States) were basically left to their own devices when faced with complex rescue emergencies. *Improvise* was often the operative word to describe operations to rescue workers buried in trench collapses, children being swept down flood-control channels, hikers stranded on cliffs, and people trapped in collapsed buildings.

Much of that began changing after a series of major El Niño-related floods swept southern California in 1982 and 1983, nearly downing several LACoFD firefighters conducting rescues in flood-control channels, natural rivers, and the Pacific Ocean. The department embarked on a program to develop swift-water rescue capabilities for first responders, helicopters, and other units. The swift-water effort was mirrored by a simultaneous rope rescue program, which trained L.A. County firefighters to manage both of these types of incidents.

After the deadly Whittier earthquake collapsed a number of large buildings in 1987, the L.A. County board of supervisors reaffirmed the fire department's role as the primary SAR agency during disasters. Fire Chief Michael Freeman ordered his staff to establish a departmental USAR committee to recommend development of a new, comprehensive system to manage USAR incidents. The USAR committee consisted of firefighters certified as California state instructors for Rescue System I, Rescue System II, Emergency Trench Rescue,

River and Flood Rescue, and other courses developed in California. In Figure 3–27, USAR firefighters and LACoFD air operations personnel practice marine disaster rescue operations.

Fig. 3–27 Practicing Marine Disaster Rescue Operations

Since that time, USAR committee members—many of whom are now assigned to the department's USAR units—have also been trained in confined-space rescue, advanced high-angle rescue, mine and tunnel rescue, helicopter technical rescue, rescue diving, helo/swift-water rescue, helo/high-rise operations, firefighter safety and survival, marine disaster rescue, ice and snow rescue, and other specialties. Several of these firefighters have been involved in FEMA's USAR program development since its inception, and all USAR committee members are assigned to the LACoFD's FEMA USAR task force (California USAR Task Force 2), as well as the USAID/OFDA/LACoFD internationally-deployable USAR task force.

Development of LACoFD's multi-tiered USAR response program: 1987 to 2000

In the early years of the program, the USAR committee was tasked with research and development, planning, development of training materials and standard operating procedures, coordination of department-wide USAR training, training of the LACoFD's USAR units, and daily administration and maintenance of the department's OES/FEMA USAR task force. In 1988, the committee recommended a plan for effective management of future technical rescues and disasters. The plan was approved by the administration that year and budgeting followed close behind. In addition to other key components, the new USAR system included a three-tiered USAR system consisting of the following resources:

First tier:
USAR first responders

All LACoFD field personnel (2,800 firefighters in 1988) were trained as USAR first responders. This included a minimum of 34 hours of training in USAR first responder topics such as structural evaluation, shoring, moving and lifting heavy objects, high-angle rescue, swiftwater rescue, and confined-space rescue, followed by periodic continuing education. This meant that every firefighter assigned to an engine company, truck company, paramedic squad, wildland patrol unit, or EST was trained as a USAR first responder. Today this training has been expanded to include Rescue Systems I, which is conducted during the recruit-training academy.

Second tier:
USAR Level 2 units

Nine existing LACoFD truck companies and the department's four helicopter units, strategically located across L.A. County, were designated as USAR Level 2 units, similar to the FDNY's squad companies. All personnel assigned to these companies were trained in Rescue Systems I, high-angle rescue, emergency trench rescue, river and flood rescue, and

other USAR disciplines. They were required to complete the USAR Level 2 training after successfully transferring to these units. The formal training was conducted by the LACoFD at newly developed USAR training facilities.

Although the LACoFD eventually phased out the USAR truck company concept in favor of expanding the number of USAR companies and task forces, it remains a valid option for many fire departments where it's impractical (or fiscally impossible) to staff rescue companies, USAR companies, and other specialized rescue units. This change allows fire departments to designate existing truck companies (already responsible for certain rescue-related duties) as specialized rescue units, essentially making them dual-role firefighters. The OCFA is one of several fire departments that have adopted the USAR/rescue truck company model with great success.

The only caveat to the USAR/rescue truck company concept is that a successful system requires the following basic protocols and policies, at a minimum:

- Mandatory qualified relief procedures to ensure 24/7 staffing of the USAR/ rescue truck companies by qualified USAR-trained personnel

- A mandatory two-year (or longer) commitment once personnel are assigned to ensure capitalization of the department's investment in training and experience

- Sufficient free time and organizational latitude to allow frequent continuing education and USAR/rescue exercises to ensure a high level of skill and familiarity

- Identification of the USAR/rescue truck companies as a specialized rescue resource

- Other considerations that are afforded to specialized units

As for the USAR-trained helicopters, they remain part of the LACoFD system today, only now they are enhanced by more advanced equipment, training, and methods. Whereas one-half of the fleet was single-engine, now it consists of two twin-engine Firehawks (the fire/ rescue version of the military Blackhawk), four twin-engine Bell 412s, and one single-engine Long Ranger outfitted as a command-and-control helicopter to help assess and manage major fires and disasters from the air. This gives the department the ability to operate with an added margin of safety over densely populated areas, hovering over high-rise buildings, conducting hoist rescues in the rugged mountains of the region, and over the ocean. These copters are well suited for transporting USAR companies and task force personnel and their equipment to distant rescue incidents—greatly reducing response times to many incidents—and to technical rescues in the mountains or the ocean where the USAR personnel augment the air operations crews for certain types of incidents. In Figure 3–28, USAR company firefighters and air ops practice helicopter high-rise team operations for firefighting and rescue operations after deployment onto the roof of burning high-rise buildings in the Marina Del Rey area of L.A. County.

Fig. 3–28 Practice Helicopter High-rise Team Operations

The LACoFD helicopters remain equipped for—and staffed with firefighters trained in—urban search and technical rescue operations, including helo/swift-water rescue, helo-high-rise firefighting and rescue operations, advanced rope rescue, hoist rescue, short-haul rescue operations, marine rescue, ice and snow rescue, large animal rescue, and other operations related to the use of helicopters for rescue. All of the air operations firefighters are certified paramedics whose medical skills and equipment lend themselves to the rescue environment. The air operations pilots are some of the most highly experienced rescue pilots in the world.

As a group, LACoFD Air Ops personnel represent part of the new wave of rescue. Some of them have invented new rescue tools and devices; others have helped develop new helicopter rescue techniques and methods, and some teach helicopter rescue. The group also includes pilots who fly rescue missions in places like Denali National Park and other high mountain locations in their spare time. They are closely aligned with the USAR companies and USAR task forces, sharing training, research, development and planning, and responding together on technical rescues across L.A. County.

Division USAR trailer caches

Early on, it was recognized that shoring materials are often in short supply in the immediate aftermath of earthquakes and other large disasters. A lack of shoring materials can complicate an already bad situation by slowing the progress of firefighters and other rescuers attempting to penetrate damaged structures to locate and extract trapped victims. To support massive shoring and digging operations after a large earthquake or other collapse-related emergency, each of the LACoFD's nine divisions has been issued a large trailer outfitted with timber, screw jacks, digging bars, rollers, pipe jacks, and other items to support convergent responders assisting firefighters at a major collapse site (Fig. 3–29).

Fig. 3–29 Division USAR Trailer

The USAR trailers can be towed to the site by a wide variety of fire department vehicles. At least one USAR trailer is dispatched to structural collapses, trench rescues, excavation collapses, tunnel/mine collapses, major traffic collisions, landslides, mudslides, and other events where cribbing and shoring are required. The ICs and USAR captains have the option of requesting as many USAR trailers as necessary to address the shoring needs of an incident.

Third tier: USAR companies

The nine USAR truck companies and four helicopter units provided an effective network of USAR service across L.A. County, but it was recognized from the outset that there was a need for specialized USAR companies or task forces with advanced levels of training, equipment, and experience to support first

responders and USAR truck companies and helicopters. These units were also needed to conduct advanced research, development, and testing; to act as USAR instructors; to function as technical advisors to ICs; to supervise rescue operations; to participate in rescue and disaster planning; and to conduct the most dangerous or technical rescues. To meet these needs, in 1990, the LACoFD created its first USAR company (identified as USAR-1) that was strategically located at the department's Special Operations Bureau headquarters, co-housed with the department's air operations heliport, its heavy equipment and transportation section, and other specialized units (Fig. 3–30).

Fig. 3–30 LACoFD USAR Company, 1991

From that location, co-housed with their air operations counterparts, the personnel assigned to USAR-1 could train daily with the helicopter personnel and respond with them to reduce response times or to assist air operations conducting helicopter-based technical rescues. From 1990 to 2000 (when the USAR company was expanded and re-staffed to create two separate six-member USAR task forces), USAR-1 was dispatched to all USAR/ technical rescues and all multi-alarm fires across the LACoFD's 2,300-square-mile jurisdiction. USAR-1 provided advanced support to the first two tiers of USAR response, provided an extra measure of safety at USAR operations, and facilitated rapid intervention operations on the fireground and at USAR emergencies.

Unlike staffing protocol for many fire departments (especially those in the east), LACoFD labor-management agreements require assignment to USAR units and other specialized companies on the basis of qualifications for the position and seniority in rank exclusively, with no provision for company officers to hand-select members. In the case of the USAR program, the LACoFD's administration and the firefighters' union agreed on a list of prerequisite training courses that firefighters and officers must complete on their own time (taught by the department) prior to submitting a transfer request for a USAR company.

Upon receiving a successful transfer to the USAR company, personnel are then required to successfully complete a lengthy battery of additional USAR training courses to raise their skills to the minimum level of preparedness to operate as a USAR company member. From that point on, it's the USAR company captains (guided by the technical operations battalion chief and supported by the department's administration) who conduct daily training to improve the skills and experience of their crews. This training is augmented by a battery of annual continuing education courses required of all USAR company personnel to remain certified and in the USAR company.

The aforementioned integration of USAR-1 with air operations yielded many positive results. The helicopter crews and USAR company crews trained extensively together. The team concept facilitated quicker, safer, more effective technical rescue operations, particularly at mountain rescues and other high-risk technical

rescue incidents. It enabled immediate helicopter response of USAR-1's crew to distant incidents and augmented the existing USAR-trained firefighter/paramedic helicopter crews during helicopter hoist rescues, short hauls, and other helicopter operations. The flying USAR company concept was instrumental in safely completing a number of highly technical rescue operations that otherwise might have taken hours longer.

When the Northridge quake struck, USAR-1 was dispatched to assist the LAFD with the rescue of Salvador Pena at the collapsed Northridge Fashion Center parking structure. Meanwhile, other USAR-trained firefighters were immediately airborne, performing damage survey missions in five of the department's rescue helicopters. As off-duty USAR personnel reported for duty as per the LACoFD's earthquake plan, they were assigned to staff additional helicopters and ground-based USAR units, including the department's FEMA USAR task force. The additional airborne USAR companies expedited the process of accounting for potential dam failures and collapse rescue situations across the affected portions of the county within about two hours. Due to severe road and freeway damage and numerous fire and rescue incidents, this survey would have taken several more hours to complete by ground alone.

USAR-1 functioned as the central USAR unit of the LACoFD until the department transitioned the USAR company into a six-person USAR task force configuration in February 2000. As of this writing, the LACoFD operates two USAR task force stations, each consisting of one USAR company and one USAR engine company.

Figure 3–31 represents both the old and the new. An LACoFD USAR Company 103 engine (1991 model) is on the left, and a 2002 model engine is on the right. Despite the difference in size, the newer (larger) fifth-wheel design has certain advantages in terms of turning radius and the amount of equipment and personnel that it can transport. The older (smaller) apparatus was capable of reaching areas where streets and roads are tighter and more curved as well as dirt motorways in some mountainous areas, but it had less overall room for equipment and personnel.

Fig. 3–31 Old and New

LACoFD Management Training for Rescue Operations

One of the key elements of the LACoFD USAR program has been a years-long campaign to familiarize all the chief officers, company officers, dispatchers, and the three incident management teams with the USAR system and its specialized resources, capabilities, and limitations. The department routinely provides management training specific to technical rescue incidents and rescue-related disaster incident management, such as response to major earthquakes, floods, and terrorist attacks.

As of this writing, no less than five dozen LACoFD chief officers (up to the level of chief deputy of department) have completed various levels of USAR training, and several of them have previously been assigned as captains, engineers, or firefighters to USAR companies, USAR truck companies, or the department's OFDA and FEMA USAR task force. This attention to training and preparedness for the command staff has a huge effect on the command of major rescues and rescue-related disasters. The chief officers running the incident are much more familiar with the breadth and scope of rescue options, and they are better prepared to select the most appropriate tactics and strategies to locate and rescue trapped victims and prevent unnecessary loss of life among firefighters and other rescuers. When rescue-related disasters such as terrorist attacks, earthquake, and floods strike, the command officers are more familiar with the regional, state, and federal rescue-related resources that may be brought to bear against the consequences of the event, and they know how to best use them.

The fire chief of the LACoFD, who currently sits on several national boards and committees related directly or indirectly to USAR and terrorism, has consistently demonstrated his enthusiasm and support for the tenets of rescue by personally participating in a string of swift-water rescue exercises, helicopter rescue demonstrations, helo-swift-water exercises, and other USAR-related activities. He has taken the time and effort to gain a firsthand appreciation for the capabilities and limitations of the department's USAR resources. It's an example of how a fire chief and his staff can positively influence the way rescue is carried out when they truly support the concept.

Swift-water Rescue Teams

The LACoFD's swift-water program began in 1982 when the department started equipping all its engine companies, truck companies, paramedic squads, and wildland fire patrols with swift-water PPE and shore-based rescue equipment, and training all its firefighters to the first responder awareness level.

In 1987, all the department's USAR truck company personnel were trained to the level of Swift-water Rescue Technician I or its equivalent River and Flood Rescue Technician I. The USAR truck companies and helicopters were equipped with more advanced swift-water rescue equipment, including gear to make contact rescues in fast-moving water and floods. In 1990, the first USAR company was staffed with River and Flood Rescue Technician Is and IIs, and equipped as a swift-water rescue team, including IRBs and other advanced equipment. Its personnel began experimenting with PWC for swift-water rescue alongside pioneers from other progressive fire/rescue agencies. This was the first full-time swift-water rescue team in southern California (Fig. 3–32).

Fig. 3–32 USAR Firefighters Practicing Boat-Based Swiftwater Rescue Methods

During the deadly floods of 1992, the LACoFD expanded its swift-water rescue net by staffing up to 11 specially trained, ground-based and helicopter-based swift-water rescue teams strategically located across L.A. County. This was the first time that a southern California fire department had established a network of formally trained and equipped swift-water rescue teams assigned to ground units and helicopters. Today, 10 of these teams are staffed by firefighters who are normally assigned to USAR task force stations augmented by other career firefighters who have taken their own time to complete swift-water, USAR, and helo/swift-water rescue certifications. These firefighters maintain their certifications and competency through the LACoFD's continuing education programs. One of the teams (assigned to Malibu) is staffed by personnel from the LACoFD Lifeguard Division, which makes good use of the expertise and experience of the department's beach lifeguards.

The USAR companies/task forces and helicopters are staffed and equipped as swift-water rescue teams 24/7. But when major storms blow into southern California, it often becomes necessary to reinforce them because of the distances to be covered and the increase in swift-water rescue emergencies in the mountains, foothills, valleys, and basins. Based on a carefully developed algorithm, these specially trained firefighters and officers are brought in from off duty to staff the additional swift-water rescue teams whenever heavy storm conditions are anticipated. The swift-water rescue teams are staffed 24 hours a day until the danger of floods and swift water diminishes to normal levels, as determined by the LACoFD's administration.

The swift-water teams assigned to the department's fire/rescue helicopters consist of two USAR firefighters certified in swift-water rescue who are specially trained to perform several helicopter-based, swift-water (helo-swift-water)

rescue evolutions. They augment the two USAR/ paramedic firefighters who are permanently assigned to each fire/rescue helicopter. This provides a pilot and a four-person rescue team prepared to fly to swift-water rescue incidents anywhere in the county within minutes (weather permitting). They will either augment ground-based teams, or handle rescues in areas inaccessible to ground-based units.

This system of swift-water rescue teams was enhanced in 1993 when a multi-agency task group designated by the county of L.A. and headed by the LACoFD's deputy chief of operations (later chief of the special operations bureau) developed a plan for other agencies to deploy similar swift-water rescue teams throughout the county. The task group also created an integrated response plan that would assure rapid response by highly trained teams when swift-water incidents occur.

USAR Task Force Fire Stations Open in 2000

In January 2000, to augment the USAR company concept, the LACoFD established two USAR task force fire stations, each consisting of one three-member USAR company and one three-member USAR engine company, for a total of six USAR firefighters per task force, all of whom are trained to exceed the standards for NFPA 1670 (*Standard on Technical Rescue Operations*) technician level. This meets the California state standard for a Type 1 (heavy) USAR company (six firefighters trained and equipped to perform rescue in the most complex and high-risk situations, including SAR operations in collapsed reinforced concrete structures, etc.).

As of this writing, USAR Task Force 103 is assigned to a fire station in the L.A. Basin and has responsibility to respond to all technical rescues and multi-alarm fires in an area that stretches across the entire southern half of L.A. County, about 1,400 square miles. Meanwhile, USAR Task Force 134 has been assigned to a fire station on the desert side of the San Gabriel Mountains (the north side of the San Andreas fault). Task Force 134 is the primary USAR unit for the northern half of L.A. County, an area of about 1,600 square miles.

In addition to meeting the California standard for Type 1 (heavy) USAR company,[4] USAR Task Forces 103 (Fig. 3–33) and 134 are trained, equipped, and staffed as Type II California swift-water rescue teams at all times. They are ready to put rescue swimmers, IRBs, and even rescue-equipped personal watercraft into flood-control channels, flash flood situations, and other swift-water and flood rescue situations. Each USAR task force is capable of being deployed via helicopter, and the crews are equipped and trained to operate in, around, and beneath the LACoFD's helicopters. They are certified to conduct the LACoFD's swift-water-helo evolutions to pluck victims from fast-moving flood waters while dangling beneath the copters on a unique short-haul rope system operated by the copters' specially trained firefighter crewpersons. They are trained as rescue swimmers for marine disasters, helo-high-rise teams, mine and tunnel rescue teams, confined-space rescue teams, dive first responders, collapse rescue teams, and rapid intervention teams. One USAR task force is dispatched to every multi-alarm fire across the county of L.A. and generally assigned as a rapid intervention crew (RIC).

Fig. 3–33 USAR Task Force 103

Since most members of Task Forces 103 and 134 are also assigned to the department's OFDA (international) and FEMA (domestic) USAR task force, they are among the most experienced in disaster and collapse SAR operations, as well as the daily technical rescues that occur across L.A. County.

Firefighters and officers interested in transferring to the USAR task force fire stations are required by the LACoFD to comply with the prerequisite USAR training standards prior to submitting transfer requests. If they receive a successful transfer to a USAR task force fire station, they are required to comply with NFPA 1670 requirements and with a number of other LACoFD USAR training and skills maintenance requirements.

Advantages of USAR Task Forces

Although national consensus seems to indicate that fully staffed USAR companies and rescue companies (preferably assigned to a special operations segment of fire/rescue agencies) are the most effective way to ensure

constant readiness to manage the most difficult technical rescues, another option is the combination of a USAR/rescue company and a specially trained engine company or truck company to create a USAR/rescue task force. This was the approach used by the LACoFD in the year 2000.

Similar to the LACoFD's hazmat task forces, which have proven effective in handling specialized operations *and* managing district responsibilities, USAR task forces have become effective tools for ICs managing the myriad technical rescues and USAR-related disasters endemic to L.A. County's vast and unique jurisdiction. The USAR task forces are capable of operating as independent companies when necessary, which lends itself to the unique training and preparedness activities required of these units. It also ensures the ability to manage the daily jurisdictional and administrative responsibilities of the fire stations to which the USAR task forces are assigned.

LACoFD personnel benefit from improved safety-related capabilities during emergency incidents, a goal that has always been at the forefront of the USAR program's mission. With one USAR task force station located on either side of the San Gabriel Mountains, LACoFD firefighters (and the public) benefit from more timely USAR support during the course of high-risk technical rescue and fireground operations. ICs have more rapid access to the experience and technical skills of personnel assigned to the USAR task forces, which are trained to assist with hazard evaluation and mitigation, incident planning, incident safety plan development, and other risk management functions.

USAR task forces may be assigned to reinforce RICs on the fireground and other emergencies, as well as performing special functions. These special functions may include structural evaluation and stabilization; heavy breaching, cutting, and lifting; assisting incident safety officers; locating hidden fire; etc. And of course the USAR task forces are prepared to conduct or assist in the full range of technical rescue operations and disasters that occur within the department's 2,000-square-mile jurisdiction.

In general, USAR task forces have brought a more diverse level of technical expertise and equipment to bear on efforts to locate and rescue any personnel who become trapped, missing, or injured during rescue and/or fireground operations. They are used in active risk management roles, as well as being primary rescuers.

Locating one Type I (heavy) USAR company on either side of the San Gabriel Mountains drastically reduced most USAR response times and provided the required redundancy for high-risk USAR operations. The plan minimizes the time required to locate, make access to, and rescue trapped victims. It expedites the response of heavy lifting, breaching, cutting, technical search, and other technical rescue capabilities to the scene when people are trapped.

The USAR task forces enhance the LACoFD's ability to conduct field training between USAR units, hazmat task forces, truck companies, and other units. It provides another resource for managing daily technical rescues, earthquakes, floods, and other disasters. The task forces also enable the LACoFD to provide immediate USAR mutual aid to surrounding agencies without depleting the rest of its jurisdictional USAR resources, an important factor in disaster-prone southern California.

USAR Task Forces 103 and 134 operate much like the LACoFD's hazmat task forces, each combining an existing engine company with a specialized company in task force configuration, yet allowing each to operate independently when necessary. Like the hazmat task forces, USAR engines and USAR companies share some jurisdictional responsibilities for fire station maintenance, pre-fire attack planning, fire prevention, and certain other administrative duties.

The USAR engines are responsible for daily emergencies within their first-in districts, reinforced by the USAR companies. USAR engines are *first up* for EMS, vehicle fires, and other single-engine responses within their districts. The USAR companies will generally be second up for EMS and other emergencies for which they are equipped.

Mountain Rescues

The Angeles National Forest in L.A. County is one of the most heavily used parks in the nation, and easily exceeds most wilderness areas for the number of rescues that occur each day. Many people visiting L.A. are surprised at the size and height of the mountains (over 10,000 ft) that traverse the region. These ranges are some of the steepest on earth, and many cities are built to the absolute limits of acceptable incline for construction. For many people in the urban wildland interface, back yards end at the boundary of a national forest. The San Gabriel Mountains are part of the USAR task forces' primary response areas and a common site for LACoFD firefighters to be confronted by difficult mountain rescues that typically last for hours In Figure 3–34, firefighters at the scene of a vehicle over-the-side with a driver trapped beneath a car, with fire that spread to the brush and unfortunately incinerated the driver alive. This emphasizes need for firefighting capabilities during many technical rescue operations, and the multiple life-threatening hazards that can confront rescuers when they arrive.

Fig. 3–34 Training for Multiple Life-threatening Hazards

The Santa Monica Mountains, which separate L.A. County's Malibu coast from the inland valleys, are steep, high, rugged, and prone to disastrous wildfires, floods, mud and debris flows, and landslides. Sitting as they do in the figurative backyard of several million people, the Santa Monica Mountains are the scene of dozens of technical rescue operations every year. These operations often involve vehicles over-the-side of mountain roads and into deep canyons; people trapped on cliffs, rocks, and waterfalls; people injured in falls from cliffs and mountainsides; airplane and helicopter crashes; rock slides; people stranded or swept away by fast-flowing creeks and streams; people trapped in deep shafts and mines; and people trapped in boats on the rocks and other similar predicaments where

the mountains meet the ocean in steep cliffs that sometimes exceed 100 ft in height.

Catalina Island is extremely rugged and mountainous (its airport is known as the *airport in the sky*) and prone to the same variety of mountain rescues that occur on the mainland, except avalanches and other snow-related mishaps because the island's mountains are not high enough to receive large amounts of snow. There are other mountain ranges in L.A. County, including the Santa Susana Mountains, Puente Hills, Hollywood Hills, and the mountains within the Los Padres National Forest in western L.A. County and neighboring Ventura County. These places are the site of dozens of mountain and high-angle rescue operations every year. As Figure 3–35 illustrates, a wide range of missions requires an equally wide range of equipment, training, skills, and experience.

Fig. 3–35 USAR Company Members Display a Wide Variety of PPE Required to Perform their Job

Deep within the rugged and heavily wooded canyons of the mountains are a multitude of isolated homes vulnerable to fire, floods, mudslides, and mud and debris flows. Danger to these neighborhoods is exacerbated after fires blow through because the denuded soils cause runoff to increase exponentially. This was demonstrated dramatically in the aftermath of the 1993 firestorms, when even moderate storms caused massive mud and debris flows and flooding. In February of that year, two hikers were killed by a fast-moving debris flow that wiped out an entire canyon after a brief thunderstorm unleashed hundreds of tons of mud from burned slopes. The mountains of L.A. County have a storied past that includes disasters that have taken hundreds of lives in the past decades and continue to present an imminent hazard to those who live in and below them (and those who venture into, through, or over them).

Vehicle Over-the-Side Rescues

Many twisting roads are built over the mountain ranges and foothills. Near-vertical drops between 300 and 1,000 ft are common alongside mountain roads. Each week, several vehicles drive over the side. These rescues often occur at night or during inclement weather (including snow), increasing the danger to rescuers. Vehicle over-the-side rescues often require advanced rope techniques just to reach the crash site. Extrication equipment, which may include hydraulic rescue systems, patient litters, and medical equipment must be lowered down as well.

Most firefighters are familiar with the difficulties of vehicle extrication under ideal conditions; that is, on flat ground, in open areas with daylight and good weather, with plenty of manpower, and with an engine company and hose lines standing by. Now consider the same

extrication in less than ideal conditions, for example, 1,000 ft down a steep cliff with the vehicle mid-slope on broken, slippery shale and a 200-ft drop to the bottom. Even worse, the accident could occur at night with rain or snow falling, with only a few rescuers available, and with a difficult hose lay to provide a protector line. Under such conditions, the danger and difficulty of extrication operations increases dramatically. In Figure 3–36, the LACoFD and Sheriff's Department units assist the U.S. Forest Service at a vehicle over-the-side that caused a brush fire and resulted in a fatality in the San Gabriel Mountains.

Fig. 3–36 At a Vehicle Over-the-side Accident with a Fatal Fire 400 ft below a Mountain Road

Figure 3–37 shows a litter team consisting of firefighters and a member of the local volunteer SAR team being lowered several hundred feet over the side to a car resting at the bottom of the canyon.

Fig. 3–37 Litter Team Rescue in Canyon

In Figure 3–38, rescuers prepare to lift a vehicle off a victim who was tossed out as the car tumbled down the mountain and burst into flames. Tragically, the victim was pinned alive and remained conscious calling for help for nearly 10 minutes as citizens attempted to reach him and assist before the flames reached and eventually consumed the victim. The accident started a brush fire requiring the response of the LACoFD and the U.S. Forest Service firefighting units.

Fig. 3–38 Firefighters Using Come-alongs and Ropes to Lift a Car off Victim Trapped in an Over-the-side Accident with Fire

Figure 3–39 shows the author during operations to lift the car off a victim during recovery operations in the canyon. Vehicles over-the-side are among the most common technical rescues in L.A. County, accounting for up to a dozen responses on busy weekends. As these photos attest, rugged terrain, steep mountain roads, restriction of helicopter access due to deep V-shaped canyons or power lines and other hazards such as fires, rock slides, inclement weather, and darkness make vehicles over-the-side a significant and challenging rescue hazard.

Fig. 3–39 Author Supervising the Recovery Operation

A patient who has been extricated from a vehicle must be properly secured in a litter and the litter properly attached to the raising system with rescuers attached as necessary. Rope rescue systems, helicopter hoist rescue operations, and other methods may be necessary to lift victims and rescuers out of the canyon.

The challenges of vehicle over-the-side rescues can be illustrated by the following true example that occurred in Malibu. A car being driven at a high rate of speed was observed literally flying some 300 feet over the side of a rugged and twisting Malibu Canyon road where there were vertical drops with heavy brush and trees growing out of the canyon walls. The driver was ejected and thrown onto a ledge with a tree branch impaled through his head, actually staking his head to the steep slope. The accident occurred at night, and it took two hours just to find the victim. Packaging the victim in the litter required the branch to be cut 3 in. from his forehead by personnel (including this author) from USAR-1 using a large bolt cutter. Pruning shears would have worked better rescuers later admitted, but the bolt cutter did the job. Rope rescue systems were used to raise the victim and three rescuers up the steep slope and through the trees and brush. The entire incident lasted four hours, typical for such rescues. The man survived his injuries after undergoing brain surgery.

Cliffhangers and Other High-Angle Rescues

The rugged mountains are alluring to hikers, rock climbers, campers, back packers, mountain bikers, and cave/mine explorers. The mountains are heavily used for these recreation purposes year-round. Several times a week (particularly during summer), people can find themselves stranded on cliffs, ledges, waterfalls, and other unusual high-angle predicaments. The mountains are scattered with old mines and caves, and explorers may find themselves trapped in various high-angle situations. Cliffhanger-type rescues often require advanced rope rescue systems, climbing techniques, and helicopter rescue methods. Sometimes they require all three.

Aircraft mishaps in the mountains

The skies above L.A. are some of the most crowded anywhere. The San Gabriel Mountains and the other ranges are notorious for plane, helicopter, and hang glider crashes. These incidents often occur under the worst conditions, for example, away from roads, on high steep faces, at night, or in poor weather, etc.). If weather conditions ground rescue helicopters, rescuers may be forced to hike in for miles. For the above reasons, plane crashes often combine all the dangers found in the course of vehicle over-the-side *and* cliffhanger rescues.

Although fatalities are common when aircraft slam into the mountains, survivors are sometimes found. For example, in 1991, four people aboard a light plane survived a spectacular crash into the middle of the San Gabriel Mountains as darkness approached. The plane broke apart as it cartwheeled across a ridge and came to rest at the top of an extremely steep wall, more than 500 ft above the bottom of the canyon.

The pilot was the only person able to leave the plane and walk. He climbed down the cliff and began walking through the wilderness in darkness. After five hours, he came upon a desolate mountain ranch and called 911. The resulting rescue lasted until 5:00 a.m. the next morning and required more than 100 rescuers from the LACoFD, the Sheriff's Department, and the U.S. Forest Service. Extreme heights and steepness required rescuers to establish a series of rope rescue systems to lower the victims into the canyon for evacuation.

River and flood rescues

The mountainous terrain creates yet another hazard: river and flood rescues. The steep terrain and high mountains act to funnel tremendous amounts of water at a high rate of speed through rural canyons and across densely populated plains to the ocean. Many of the natural streambeds resemble the wildest of rivers during rainstorms. The endless cycle of wildfires and flooding are well documented, and few situations are more dangerous than attempting the rescue of a victim stranded at the mouth of a steep canyon as flood waves of water, mud, and debris rush down fire-denuded slopes. This scenario has played itself out many times in the mountainous areas of L.A. County over the years.

Another major swift-water hazard is found where concrete-lined flood channels flow through densely populated areas. Some of these channels are 40 ft deep, up to 600 ft wide, flow at speeds exceeding 30 to 40 miles per hour, and sometimes dive under the ground for miles. Victims being carried by water moving at such velocities can be swept for miles before rescue operations can be organized and conducted. Under such conditions, swimming ability and experience making contact rescue in calm water or oceans has little bearing on the ability of rescuers to save the victim.

Since 1982, the LACoFD has operated a formal system of swift-water rescue, using special training, equipment, and teams. Successful operations in swift-water rescue situations depend upon sufficient resources to handle essential tasks. Some responses in L.A. County flood-control channels involve immediate response of 60 to 70 firefighters and other rescuers. These responders are trained personnel who can quickly establish technical rope systems. Often, they have boat-handling skills for high-line system boat rescues. Generally, they are personnel with good skills and basic equipment for attempting shore-based rescue. These operations require good planning and management procedures to assure responding resources

are placed where they will do the most good and the ability to make proper tactical and strategic decisions based on good training and experience.

Today the LACoFD uses its rescue helicopters extensively for swift-water rescue operations and has developed trailblazing new helicopter-based, short-haul rescue methods for victims being swept down flood channels.

A county-wide system of swift-water rescue teams and integrated response has been developed in cooperation with other fire/rescue agencies. That system has expanded across southern California and now includes dozens of swift-water rescue teams that are inventoried and available for mutual aid responses to swift-water emergencies and disasters anywhere in California (see appendix I).

Surf and ocean rescues

The Pacific Ocean creates another significant hazard for firefighters and other rescuers. The lifeguard division of the LACoFD provides rescue service along most beaches, including Catalina Island. There are times, however, when firefighters arrive on scene first, particularly at night, during winter, and in certain geographical areas where fire stations are close to the beaches (Fig. 3–40).

Fig. 3–40 L.A. County FD Personnel Practicing Cliff Rescue Operations

On the Palos Verdes Peninsula, where the border between land and ocean is marked by 300-ft vertical cliffs running for miles, the fire department is frequently on scene first. Technical rope systems or helicopters are often required to reach the rescue site in a timely manner. LACoFD firefighters have saved the lives of several people by conducting contact rescues in the ocean below the cliffs of the Palos Verdes Peninsula.

Snow and ice rescues in the mountains

The 10,000-ft San Gabriel Mountains are covered in snow all winter, and this creates yet another rescue hazard for which the LACoFD must be prepared. These mountains are home to several ski areas subject to avalanches that have buried several out-of-bounds skiers in recent years. Some skiers and snowboarders have become lost, requiring days-long SAR operations and involving dozens of searchers supported by helicopters and fixed-wing aircraft.

The steepness of the San Gabriels—some of the steepest mountains in North America—make avalanches inevitable. Avalanches have swept over large stretches of the famous Angeles Crest Highway, threatening to bury automobiles or sweep them over the edge of the road into deep canyons. Less than two weeks after the Northridge earthquake, a one-quarter mile stretch of the Crest Highway was covered 20 ft deep in snow by an avalanche that struck just after an auto passed the area. The LACoFD dispatched several engines, a truck company, two USAR units, a canine search team, one firefighting bulldozer and track loader, a battalion chief, and a fire/rescue helicopter. They also requested a mountain SAR team. Several other agencies joined in an hours-long search using avalanche poles and other traditional methods, while heavy equipment gingerly assisted by moving

boulders and mounds of snow to reach areas where automobiles might be buried. Fortunately, every car on that part of the road managed to miss being buried.

In the San Gabriels, several mountain lakes freeze over and create conditions that occasionally lead to tragedy when visitors break through the ice while tobogganing or engaging in other forms of snow play. In February 1993, three people perished when they went through the ice over a lake in the San Gabriels despite the efforts of firefighters, U.S. Forest service personnel, and SAR teams who managed to rescue several *would-be* rescuers. Other fatalities have occurred in icy mountain lake waters in recent years. The LACoFD USAR task forces and air operations units are trained and equipped to operate in these environments and to conduct rescue operations in conjunction with the U.S. Forest Service and the L.A. County Sheriff's Department whenever people find themselves in trouble in the mountains above L.A.

Man-made rescue hazards

Naturally, the vast majority of LACoFD fire stations are located in some of the busiest urban areas in the nation. As in most densely populated urban areas, many of the technical rescues that occur are related to man-made causes. These include vehicle extrication, trench and excavation collapses, structure collapse, stranded window washers, workers trapped in machinery, people trapped beneath industrial materials, train wrecks, and plane crashes. Traffic collisions can involve semi-trailer trucks, cranes, dump trucks, motor homes, and other large vehicles. Sometimes, jumpers threaten to throw themselves (or others) off buildings, bridges, and other high points. Incidents can also involve confined-space rescues, people swept downstream in flood-control channels, explosions, and terrorist attacks. Occasionally, SWAT operations require technical search equipment and special extrication tools. Animals also become trapped in various predicaments. The multi-tiered USAR response system employed by the LACoFD has proven very effective for these incidents over the years.

Dam failure

Another man-made hazard that can spawn a disaster is the more than 200 dams and flood basins that have been built across L.A. County. Several of these dams hold back more than 50,000 acre-ft of water and are located upstream of densely populated areas. Combined with the local seismology, these dams create a major hazard under adverse conditions.

The 1971 Sylmar earthquake severely damaged the Van Norman Dam, causing emergency evacuation of thousands of people in the San Fernando Valley. The largest life loss ever in California resulted from the 1928 collapse of the St. Francis Dam, which swept more than 500 people to their deaths in L.A. and Ventura Counties.

Marine mishaps and disasters

In the crowded Pacific waters off L.A. County, boats capsize, cargo ships collide, cruise ships catch fire, airplanes (including at least three airliners in recent years) crash into the ocean, windsurfers get lost in the 26-mile wide Catalina Channel, and sometimes murders are staged as accidents aboard boats and yachts. There was even a case of a tugboat being pulled to the bottom after its tow lines were snagged by a passing nuclear-powered submarine, killing all aboard the tug. And now there is the ever-increasing potential for terrorists to use cargo ships as WMD by planting nuclear

bombs aboard them, to rig oil tankers as huge floating bombs that can be detonated in harbor, or board cruise ships to take hostages and conduct other acts of terror. In Figure 3–41, the crew of a USAR unit saw a mini-sub being transported through L.A. and took the opportunity to plant a seed of suggestion for the future.

Fig. 3–41 New Approach to Marine Disaster Rescue Operations?

In the early 1990s, the FAA mandated the creation and testing of air-sea disaster plans for every major airport located adjacent to a large body of water. In the case of L.A. International Airport (LAX), one of the main concerns was that of airliners crashing into the ocean on takeoff. Two such events have already occurred at LAX. At least one other airliner has crashed into the ocean from high altitude. Helicopters and/or private airplanes also crash into the sea off L.A. County almost every year.

In response to the FAA mandate and to increase the level of response to non-airliner marine disasters, the U.S. Coast Guard, L.A. County Lifeguards, LACoFD, LAFD, L.A. County Sheriffs, and other allied agencies developed an elaborate air-sea disaster plan.

Part of the plan calls for both LACoFD USAR task force fire stations to board LACoFD fire/rescue helicopters to deploy into the ocean as rescue swimmers. The firefighters are trained to deploy from the copters into the ocean wearing wetsuits and other PPE with specially designed inflatable platforms and other gear. They are then expected to begin plucking people from the water while other agencies respond with boats and helicopters to help move people to a shore-based mass-casualty operation that will be automatically established as part of the marine disaster plan.

Mountain train tunnels

Several train tunnels have been bored through the San Gabriel and Santa Susana Mountains. These tunnels carry thousands of passengers each day; they also accommodate much of the freight that's brought into L.A. The Newhall Tunnel is one mile long, and its south portal exits the mountains at exactly the same spot where the elevated interchange of Interstate 5 and Highway 14 collapsed during the Sylmar earthquake in 1971 and then again 23 years later in the Northridge quake. Another mountain tunnel known as the Soledad Tunnel is one-half mile long.

On the west side of the San Fernando Valley is yet another tunnel that's been driven through a seismically created mountain range. The Chatsworth Tunnel is actually a series of three tunnels ranging from one-quarter to three-quarters of a mile long. These tunnels run through the Santa Susana Mountains that serve as the border between L.A. and Ventura counties. As fate would have it, the Chatsworth tunnels run adjacent to (and through) the very same fault system that caused the Northridge earthquake that raised the Santa

Susana Mountains 9 inches in a matter of seconds. The large number of passengers, the length of the tunnels (which require special entry procedures for fires and rescues), the potential for underground disasters related to terrorism, earthquakes, flooding, and other causes, make these mountains a special hazard.

Rail tunnels under the city

Many issues related to mountain train tunnels also apply to underground rail tunnels beneath urban areas. Urban L.A. County is bisected by a number of underground rail lines, including the Metro Rail system (completely underground) and the Alameda Corridor, a subterranean rail system without a roof, like a 30-ft deep trench open to the sky with freight trains running through it all day long.

Part of the Metro Rail system (actually two tunnels running parallel) runs from Hollywood directly beneath an entire mountain to the San Fernando Valley, a distance of about six miles that has only one stop, Universal Studios. At the apex of the hills beneath which the tunnels have been bored, trains run more than 600 ft beneath the surface, with very limited access (through vent shafts, etc.) to points between the Valley and Hollywood stations. Not surprisingly, the tunnels have been bored directly through earthquake faults, ancient underground rivers, and other geologically unstable elements. Given the potential for natural disasters and man-made ones (including the possibility of terrorism), it's a cause for concern among rescuers and firefighters.

Then there is the potential for fire. Almost all of the underground rail systems have been drilled through soil that is laden with hydrocarbons from ancient seeps, oil fields, and even the same terrain that created the famous La Brea tar pits. The tunnels are constantly monitored for intrusions of explosive, flammable, and toxic gases, and the fire systems have been hardened to address concerns caused by earthquakes and other hazards. But the potential for terrorist attacks involving explosives is one example of a situation that could conceivably render such systems useless. Lessons learned in the 1990 Metro Rail tunnel fire that resulted in the catastrophic collapse of one-quarter mile of the tunnel near downtown L.A. are being heeded as a LACoFD planned future response. For personnel assigned to the USAR, hazmat task forces, and other units that would be expected to perform firefighting and rescue in these places, safety and effectiveness are significant concerns.

Confined-space rescue

The LACoFD responds to a large number of confined-space rescue operations each year. It's among the most common types of USAR incidents to which the department responds and definitely among the most dangerous. The common perception of confined-space rescue as an industrial problem that happens mostly in urban areas is sometimes shattered by the types of confined-space situations encountered by LACoFD personnel. To be sure, the LACoFD responds to plenty of confined-space rescues in oil refineries, chemical plants, subterranean vaults and shafts, deep vertical shafts at construction sites, and the like. But some confined-space situations encountered by the LACoFD aren't found in any textbooks on rescue.

In Figure 3–42, L.A. County USAR firefighters extract a victim from the bottom of a 50-ft deep shaft in the Malibu hills. This rescue required the crew of USAR-1 to be flown to the scene in an LACoFD helicopter. One firefighter was lowered into the shaft to create access and protect the victim and a host of

USAR specialists and local LACoFD firefighters performed duties such as rigging, ventilation, air monitoring, rope teaming, medical teaming, etc. Deep shaft rescues like this are extremely dangerous and challenging, combining the hazards and requirements of confined-space rescue and high-angle rescue

Fig. 3–42 Deep Shaft Rescue Operation (Courtesy Gary Thornhill)

Examples of confined-space incidents include children trapped beneath concrete culverts undermined by years of erosion from floods that finally collapsed them and trapped the children while they were playing. Other incidents occur when hikers become trapped in the vertical shafts of towering reinforced concrete flood-measuring stations deep within the rugged gorges of the San Gabriel Mountains. Workers can become trapped in the flooded interior recesses of large dams when spill gates malfunction. Victims have also been sucked into the intake channels of power plants. One victim of an auto accident had his entire car (with him in it) carried for several miles in an overflow channel that went underground through a deep sump system where he was finally spit into a lake. He actually lived to tell about it.

Trench rescue

L.A. County is experiencing fast growth from new construction and the renovation of existing construction and civic projects that require excavation and trenching. The county is accustomed to trench and excavation collapses that trap workers and other victims. This is especially true in southern California where the geological conditions are often unstable to begin with and compounded by the effects of new or repeated construction, vibrations from traffic and other causes, and even the occasional earthquake. Like many fire departments, the LACoFD's USAR units are trained and equipped to comply with OSHA and Cal/OSHA standards for entry into trenches and excavations for rescue.

Figure 3–43 shows LACoFD USAR firefighters assisting engine and truck company personnel as they struggle to rescue a worker pinned by two large boulders that rolled into an excavation during construction of Universal Studios' *City Walk*. This rescue required four hours of intense work and the use of air

bags, screw jacks, an aerial ladder, ropes, and hand tools. These tools were used to stabilize the walls of the excavation and carefully separate the boulders from the victim so that he could be treated for crush syndrome. Finally, he was carefully extracted without causing a secondary collapse or a fatal movement of the largest boulder.

Fig. 3–43 USAR Firefighters Remove Victim from under Boulders

Just as in confined-space rescue, the unusual projects, often sites in geologically unstable terrain, combine to create unusual trench/excavation conditions. These operations typically last hours and present firefighters and rescuers with unusual hazards.

When filling firefighter or officer positions, there is no allowance for hand-selecting personnel for positions of rescue companies or other units. In these agencies, the staffing of ladder companies, squads, and rescue companies is strictly based on seniority.

In some agencies there is an added prerequisite: a minimum agreed-upon level of qualification for the position in question. Seniority is considered the *tie breaker* in case two or more qualified firefighters are vying for the same position.

In the absence of hand-selecting rescue company members, one method that's proven effective is to implement meaningful standards of training and skill that must be met and documented before candidates can become eligible for bidding positions. Personnel having the necessary minimum qualifications can then be selected based on seniority among qualified candidates. Some departments have adopted NFPA 1670 (*Standard for Fire Department Technical Rescue*) as the basis for their standards. Others have implemented local training/qualification standards, and still others have adopted combinations of NFPA 1670 and local standards. This is an acceptable compromise when labor-management agreements (documented by memorandums of understanding or agreement) include provisions that ensure the proper qualifications commensurate with the demands and dangers of the mission of the units.

Unfortunately, labor/management relations in some fire/rescue agencies are such that it's difficult to implement meaningful qualifications standards in the first place, and even more difficult to enforce them.

Training courses

There are a number of choices for the modern fire department to ensure that their rescuers are properly trained. Fire department-based rescue courses available in some areas and in some agencies may include (but are not limited to) the following:

- Rescue Systems I
- Structure Collapse Rescue Technician
- Emergency Trench Rescue
- River and Flood Rescue Technician I (or Swiftwater Rescue Technician I)
- Confined-space Rescue
- Annual confined-space continuing education
- Rescue Systems II
- Advanced Rope Rescue
- Helo/Swift-water Rescue (with annual recertification)
- River and Flood Rescue Technician II/Swiftwater Rescue Technician II
- River and flood rescue continuing education
- Helicopter hoist rescue certification and 90-day skill recertification
- Mine and tunnel rescue team certification

- Rescue swimmer operations for marine disaster (blue water) incidents
- Helicopter insertion operations for rescue swimmers
- High-rise helicopter company operations
- Ice and snow rescue operations
- Dive rescue support training
- Various technical mountaineering skills

Training courses are covered in greater detail in chapter 5 (Rescue Training).

Use of Helicopters for Rescue

To ensure timely response to emergencies—particularly when ground-based response times exceed 30 minutes—some fire/rescue agencies utilize fire/rescue helicopters to transport their rescue companies and USAR units. This is a method that has been used successfully for years by a number of fire/rescue agencies.

For example, LACoFD USAR task force fire station captains have the option of requesting helicopters to transport their crews and special equipment to distant incidents, remote mountain rescues, marine disasters, Catalina Island, and other hard-to-reach locations. For certain incidents (e.g. marine disasters, high-rise fires, mountain rescues, etc.) USAR task force members are sometimes transported by helicopter to facilitate unique deployment options based on the discretion of the IC, USAR task force captains, etc. (Fig. 3–44). This includes rescue swimmer insertion into the ocean or lakes, deployment onto the roofs of burning high-rise buildings, hoist insertion for mountain rescue, swift-water/helo operations, and other specialized tactical operations (see chapter 5 in Volume II, Helicopter Rescue Operations).

Fig. 3–44 LACoFD USAR and Air Ops Personnel Practicing Helo-swiftwater Rescue Operations

Just as rescue companies and USAR task forces may request aerial transportation to expedite or increase the effectiveness of a response, LACoFD air operations units conversely have the option of requesting USAR task force personnel to board helicopters to assist them with high-risk operations. When air operations units are dispatched to assist outside agencies with cliff rescues, vehicles over-the-side, plane crashes, and other potentially high-risk SAR operations, they may request USAR task forces as support resources. The intent of this optional capability is to ensure adequate trained personnel to conduct helicopter-based operations, as well as to ensure adequate rapid intervention capabilities.

Rescue Standards and Regulations

Some members of the fire/rescue community (especially some supervisors) consider worker safety-related standards and laws as the bane of their existence.

Others see them as the basis for adopting common sense rules to help ensure reasonable safety for firefighters and others engaged in high-risk firefighting and rescue operations. Some ICs feel that the rules tie their hands and make it more difficult to manage emergency operations. Still others are thankful for the framework of precautions that the worker safety regulations provide.

To be sure, some regulatory agencies and individual inspectors have been known to take things to the extreme, interpreting and enforcing worker safety laws beyond the realm of what some might consider common sense. Take, for example, one small fire department that was recently fined $55,000 for violations found during an inspection of fire station facilities. Most of the violations were discovered in connection with decades old hose towers. Among the problems cited were 15-in. wide ladders (regulations specify 16-in.), rungs of the hose tower ladders 2 in. too far apart, platforms too narrow at the top of the towers, and no raised edges on the platforms to prevent accidental slips and falls. At the time of this writing, the fire department in question is in the process of correcting the violations but is protesting the fine.

Some might argue that the inspector went overboard because the hose towers were built before current laws went into effect, and similar violations can probably be found at fire stations across the nation. In essence, there are probably more pressing safety problems in the fire service than the width of ladders on a hose tower. After all, firefighters are expected to climb and operate from ladders under far worse conditions on the fireground and at the scene of technical rescue operations. On the other hand, if worker safety agencies aren't there to ensure that life safety laws are being followed, who will ensure the safety of firefighters when it comes to more high-risk issues, such as regulations related to confined-space rescue and interior firefighting operations?

The lesson seems to be that regulatory agencies and the inspectors who represent them are a mixed blessing. To some, they represent a hindrance to getting the job done. To others, they lay down the law that says fire/rescue agency administrators—and line supervisors—must respect the sanctity of life safety for their employees whenever feasible, given the serious risks that we all voluntarily take when we put on the badge.

OSHA recently instituted laws that affect fire/rescue agencies everywhere. Not only are firefighters and other rescuers required to operate under strict guidelines for confined-space and tunnel rescues, but now they are required to meet a new standard for rapid intervention in the form of the so-called two-in, two-out rule. Civil and criminal penalties can result from failure to adhere to these and other OSHA regulations, and fire/rescue agency administrators (not to mention fire captains and other direct supervisors of firefighters injured or killed in the course of their duties) may be held personally accountable.

Several state worker safety agencies have become much more pro-active in recent years, instituting regulations even more stringent than those of OSHA and holding public agencies and their administrations accountable for violations.

California, which has its own worker safety agency (Cal/OSHA) is one of a handful of states that have enacted serious changes in the way public agencies and their representatives are dealt with. For the purposes of this book and its emphasis on rescuer safety, it's instructional to review some of the sweeping changes that have occurred in that state. California Assembly Bill 1127 (AB 1127) radically changed the California Labor Code by holding all California employers (including fire/rescue agencies) subject to significant civil and criminal penalties. The law became

effective on January 1, 2000. As a result of AB 1127, the following actions are now possible or mandatory in California:

- The state created the position of complaint investigator, who must investigate allegations of retribution (e.g. harassment or discrimination) by employees who have reported unsafe conditions to Cal/OSHA, their union, or other worker safety organizations or employee representatives.

- The bill mandates investigative action and follow-up action on serious and non-serious complaints and increases the time during which an employee may file a complaint to six months after the event.

- A Cal/OSHA citation can be submitted as evidence in any personal injury or wrongful death action.

- Cal/OSHA is now required to maintain records of all complaints against employers, including fire departments, police departments, and SAR organizations.

- The bill greatly increases the civil penalties for repeated violations of the worker safety standards and for refusing to comply or inducing others to engage in repeat violations.

- Penalties enacted by the bill may apply to the employer, the manager, or even the line supervisor (i.e. fire department captains, battalion chiefs, rescue team leaders, etc.).

- Fines against the supervisors may be up to $15,000 per violation and a maximum of one year in jail.

- The bill makes it easier for temporary workers to file civil actions against secondary employers.

- The initial base penalty for a serious penalty is $18,000 with a maximum of $25,000.

- Any company or employee having management control who willfully violates any OSHA standard that results in death or causes prolonged or permanent impairment of the body may be fined up to $250,000 and imprisoned for up to 16 months.

- A corporation may be fined up to $1,500,000.

- Multiple convictions for willful violations may result in corporate fines up to $3.5 million.

- For multiple serious/willful violations, no adjustment for good faith will be given. Violations of special orders may result in fines ranging from $5,000 to $70,000.

- Fines up to $300,000 may be levied against employers who affirm in writing the abatement of conditions when no abatement has occurred.

- If a worker, supervisor, or corporation recognizes a hazard and fails to mitigate it and serious injury or death results, it is classified as a willful violation.

- Applicable regulations and standards for rescue.

A growing body of regulations and standards from other organizations related to firefighting and rescue operations is forming a slow consolidation of the shared wisdom of disparate schools of thought on rescue into increasingly well-defined practices. As these regulations and standard practices continue to evolve, rescue will become more widely recognized as a professional discipline. For the purposes of training, these disciplinary standards can be used as guides to set the direction of rescue programs. Among others, the following regulations and standards provide the basis for developing an effective rescue training program and for determining the efficacy of its application.

- NFPA 1670, *Standard on Technical Rescue*. This standard establishes guidelines for career tracks in rescue and recognizes the importance of developing knowledge, skills, and abilities (KSAs) that are quantifiable.

- NFPA 1006, *Standard on Professional Competencies for Technical Rescue*

- NFPA 1983, *Standard on Life Safety Rope, Harnesses, and Hardware*

- NFPA 1470, *Standard on Search and Rescue Training for Structural Collapse*

- NFPA 1500, *Standard of Fire Department Occupational Safety and Health Programs*

- NFPA 1561, *Standard on Fire Department Incident Management Systems*

- OSHA 1910.134, *Respiratory Protection*

- OSHA 1926.650, *Trench and Excavations*

- OSHA 1910.146, *Permit Required Confined Spaces for General Industry*

Of these standards, two in particular deserve further clarification, NFPA 1670 and 1006.

NFPA 1670

No discussion on urban search and technical rescue training would be complete without mention of NFPA 1670, *Standard on Operations and Training for Technical Rescue Incidents*, adopted by the National Fire Protection Association (NFPA) in 1999. This national consensus standard should be considered the baseline for establishing USAR and technical rescue systems. Its origins lie with NFPA 1470, *Standard on Structural Collapse Training and Operations* and NFPA 1983 and NFPA 472 (*Standard on Hazardous Materials Operations*), which established benchmarks for fire/rescue services engaged in those operations.

The effects of NFPA 1670 on the new standards for technical rescue are far reaching. Its origin is also worth noting. While NFPA Standards 1470, and 1983 covered specific aspects of certain types of technical SAR operations, there was no all-encompassing standard for the full range of disciplines that come under the heading *Rescue*. Consequently, fire departments and other public safety agencies ran into problems on several fronts.

Because there was no nationally-recognized regulation guiding the establishment of formal rescue systems, well-meaning administrators who wanted to enhance the safety of their personnel were hamstrung by the lack of a specific standard they could cite as evidence of the need to enact meaningful requirements for rescue and USAR units. The lack of a national standard meant there was little impetus to prompt less-willing fire/rescue agencies to adopt formal rules about training, staffing, and equipment for rescue. Some labor unions protested when fire departments attempted to establish formal rules for training and staffing. Without a recognized standard upon which fire departments could base their requirements, some unions had a field day with their administrations, who soon realized that their hands were essentially tied behind their backs. In other words, there were no *teeth* to training standards for various types of rescue operations.

Without a national standard, it was up to every individual city, county, state, and training organization to establish its own requirements that often varied widely. This naturally led to significant differences in rescuer competence from one state or county to the next, between different fire/rescue agencies, and even from one unit to the next within the same agency.

As rescue evolved toward its present state, these problems required redress. A growing number of fire/rescue agencies already had units conducting a wide range of technical rescues. A few fire departments, by virtue of their location, terrain, weather, demographics, seismology, and other factors, were thrust into the position of managing very hazardous and complex rescue and disaster problems. A standard was needed to encompass the safety needs of firefighters conducting all these rescue-related operations; hence NFPA 1670 was created. What follows is a review of the main points of this trailblazing national standard for rescue.

One of the interesting features of NFPA 1670 is the *career path,* an algorithm of sorts that allows progress from awareness to first responder to operations to technician. It's a proven approach employed by a num-

ber of fire/rescue agencies across North America. As one example, since 1987 the LACoFD has required its firefighters intending to bid a USAR company to first complete a battery of rescue training that begins with the basics (Awareness) and gradually includes more comprehensive and complex courses that bring the student through the Operations level until they finally achieve the Technician standard. It's an approach that has proven itself in a long string of successful rescue operations during disasters and complex technical SAR emergencies during the past decade and a half.

NFPA 1670 includes seven main disciplines that represent the bulk of fire department rescue responsibilities nationwide. Obviously there are a number of subsets of these specific disciplines based on local and regional conditions. And these can be addressed by any individual agency. What 1670 does is lay down a framework upon which any agency can base a viable rescue program.

NFPA 1670 identifies operational levels (Awareness, Operational, and Technician) that are available to any fire/rescue agency that wishes to take up a specific rescue discipline. It also delineates the training levels required to meet each operational level.

The *Awareness* level is essentially for first responders who must—to operate effectively and with a reasonable level of personal safety—understand how to size-up, assess, and manage the rescue scene until the victim is rescued or until the arrival of Operational or Technician level units arrive. These responders should understand the basic dynamics and hazards of the rescue emergency. They should be prepared to evaluate the need for additional assistance and call for it without delay. They should be prepared to safely commit themselves to the incident, or if the danger exceeds the level of protection, to establish an exclusion zone and deny entry.[5] Finally, they should be prepared to control and manage the rescue scene.

The *Operations* level includes personnel who have met Awareness level requirements as a baseline and who have gone on to receive additional instruction and experience. This training and experience should enable them to review the size-up of the Awareness level rescuers, to determine probable or confirmed victim location and survivability, to make the rescue scene reasonably safe for emergency operations to proceed, to protect victims and package them as necessary, to use (and supervise the use of) rescue equipment, and to safely and effectively apply appropriate rescue tactics and strategies. Essentially, these are the secondary responders to rescue emergencies.

In the NFPA 1670 scheme, the *Technician* level represents the personnel with the highest level of training and experience and the most advanced equipment to supervise and conduct rescue. Technician level rescuers should meet or exceed the Awareness and Operations level requirements as a baseline. They should be prepared to assess and reassess the rescue problem and devise and appropriate strategy supported by proper tactics to solve it while maintaining a reasonable level of safety for rescuers. They should be able to provide personal protection for the victim and rescuers alike during the course of operations. These individuals should be prepared to advise the IC about specialized resources and to supervise or conduct advanced methods in complex SAR situations. In short, NFPA 1670 Technician level personnel should be the rescue experts for your agency.

Chase Sargent, in a recent *Fire Engineering* magazine article noted that "the standard also allows different geographical areas of the country to compare programs and service levels based on specific *knowledge*, *skills*, and *abilities* (KSAs) rather than program names or descriptions. Finally, people from the East Coast, West Coast, and Midwest can talk about levels of service based on KSA's instead of program names and descriptions…Organizations will be able to speak the same language and create a more efficient program evaluation nationally. This benefits all organizations when discussing reciprocity, cross-certification, and service levels, allowing them all to use common terminology…For example, FEMA USAR documents are eliminating the reference to such things as 'Rescue Systems 1 or equivalent', opting instead to provide specific KSA's as outlined in the NFPA 1670 standard." [6]

Like the other NFPA standards, 1670 is not binding, and some fire/rescue agencies may choose not to adopt it. But because it is a recognized national standard, agencies that don't adopt it (or exceed its requirements) could find it being used against them in liability lawsuits and other legal cases. Not only that, in the event of a *rescue gone bad* (especially one that's filmed by news cameras, aired live, and rebroadcast over and over) the news media and public officials could cite NFPA 1670 to argue that the agency was not adequately prepared to manage a local rescue hazard. For those who ask: "Should my agency adopt NFPA 1670 as a minimum standard?" The answer might well be another question: "Does your agency respond to rescue emergencies?" If the answer is yes, it's probably a good idea to adopt NFPA 1670 and other rescue-related standards as needed.

Selected passages from 1670

The following are some important passages from NFPA 1670:

- *Section 2-1.6*—"The Agency Having Jurisdiction (AHJ) shall provide for training in the responsibilities that are commensurate with the identified operational capability of each member."

- *Section 2-1.6.1*—"The AHJ shall provide for the necessary continuing education to maintain all requirements of the organization's identified level of capability."

- *Sections 2-5.1.1 and 2-5.1.2*—"All personnel shall receive training related to the hazards and risks associated with technical rescue operations… and shall receive training for conducting rescue operations in a safe and effective manner…"

- *Section 4-1.2*—"The AHJ shall evaluate the effects of severe weather, extreme heights, and other difficult conditions to determine whether the present training program has prepared the organization to operate safely."

- *Section A-2-5.4*—"The AHJ should address the possibility of members of the organization having physical and/or psychological disorders (e.g. fear of heights, fear of enclosed spaces) that can impair their ability to perform rescue in a specific environment."

- *Section A-3-2.2*—"The AHJ should train members to recognize the personal hazards they encounter and to use the methods needed to mitigate these hazards in order to help ensure their safety."

- Section A-4-3.2—"Rescuers should be trained to perform these procedures under the environmental (e.g., snow, darkness, wind, and elevation [e.g. potential height]) conditions."

NFPA 1006

Fire/rescue agencies should also be aware of NFPA 1006, *Standard for Professional Competence for Responders to Technical Rescue Incidents.* NFPA 1006 is a national consensus standard intended to augment NFPA 1670 by providing the framework by which fire/rescue agencies can ensure the readiness and competence of their personnel to conduct rescue operations. The following is a paragraph from NFPA 1006, Section 2.1:

Because technical rescue is inherently dangerous and rescue technicians are frequently required to perform rigorous activities in adverse conditions, regional and national safety standards shall be included in agency policies and procedures. Rescue technicians shall complete all activities in the safest possible manner and shall follow national, federal, state, provincial, and local safety standards as they apply to the rescue situation.

NFPA 1500

The effects of NFPA 1500, the *Health and Safety* standard, are equally far-reaching. This standard is representative of the new emphasis in reducing injuries and death among fire/rescue agency personnel. Gone are the days when firefighter deaths were accepted as par for the course and something to be expected, for us simply to bear. Instead, firefighter safety regulations have given us sophisticated methods to combat the tragic death tolls that plague our profession.

Case Study 2: LACoFD swift-water rescue solutions

A history of damaging storms reveals swift-water and flood hazards

We've already established that the development of a rescue system begins with assessing the local hazards and rescue potential. Sometimes the hazards are quite obvious if one looks at patterns that develop over a period of years. For example, in L.A. County there was ample evidence of the need to establish a formal swift-water rescue system, and this was the basis for the department's multi-tiered swift-water program developed in the early 1980s. There have been thousands of swift-water rescue emergencies in L.A. County during the past century and a half, and they tell a story of lethal hazards that exist every day and are amplified exponentially during major storms.

A long history of floods and swift-water disasters

Flooding is part of the history of L.A. Storms are known to have left huge areas of southern California under water since the Spanish first settled it. This region is laced with foothills and 10,000-ft mountain ranges that act as funnels, sending rainwater and snow melt racing down steep canyons. The water picks up tremendous amounts of debris as it rushes across a checkerboard mix of rural and suburban areas on its way to the ocean. In some of these areas, whole cities have been built directly on historic flood plains with little regard for the ability of nature to wash them away.

In 1938, a series of storms left most of the L.A. Basin and parts of neighboring Orange County under water and killed dozens of people. The same general inundation area is now populated by millions of people. These storms, following on the heels of previous flood disasters in L.A., prompted the U.S. Army Corps of Engineers and local flood-control districts to design a vast system of flood channels and dams to withstand a 100-year storm in L.A. County. The maximum flows for a 100-year storm were based on formulas from geologic records available at the time.

However, recent discoveries have led to the conclusion that those figures were far too conservative. For example, new calculations developed by the Corps have indicated that one deluge in 1868 was at least three times larger than the 1938 storms. Hydrologists are convinced that the stream originating in Big Tujunga Canyon, a mountain tributary of the L.A. River, flowed a nearly incomprehensible 300,000 cubic feet of water per second (cfs) during the 1968 storm. In comparison, the typical flow of the Colorado River through the Grand Canyon hovers somewhere between an average of 15,000 to 25,000 cfs.

To stand where Big Tujunga Wash meets L.A. County's Foothill Freeway, which skirts the southern boundary of the San Gabriel Mountains and separates much of the populated area from the Angeles National Forest, and imagine a flow that exceeded the Grand Canyon by a great factor is to begin to understand the dimension of disaster that may someday befall L.A. County. Then to realize that Big Tujunga

Wash is only one of many large tributaries that pump water into the L.A. River is to understand just why flood-control officials are justifiably concerned about the possible effects of so-called *great storms* like those that accompany the periodic return of El Niño in the Pacific.

The consequences of an 1868-sized storm today in this densely populated area are nearly unfathomable to a population that has been protected from smaller flows by the modern flood-control system. Today's inhabitants have not experienced anything even approaching the volume of water dropped on L.A. by the 1868 storms.

Considering the scale of geologic time, the 1868 storms were only a tick on the clock away from the present. It is not too far-fetched to imagine a similar series of storms, perhaps spawned by a major El Niño event, striking the region. Today, it is clear that a storm of similar magnitude would easily overwhelm the flood-control systems and cause massive devastation and death across the region. And there seems to be one unanswered question about the storms of 1868. Is it possible that they were simply the result of an El Niño event the size of which is *currently* affecting the Pacific Ocean?

In many areas of L.A. County, the only thing preventing floods each winter is a complex system of flood-control structures. There are hundreds of flood debris basins. Over 500 miles of concrete flood channels bisect the developed areas. Some of these waterways flow water throughout the year. However, most are dry except in the winter and spring. Under these conditions, it is easy to be lulled into complacency about the potential for swift-water rescue incidents to occur.

But the lack of year-round flow is dangerously deceptive. Those who have seen water roaring down rivers, streams, and flood channels after a storm have a healthy respect for the power of moving water. In concrete-lined flood channels, speeds can exceed 35 miles per hour, and volumes can be tremendous.

During high flows, the normally dry L.A. River carries more water than the Mississippi River. The water carries tremendous amounts of floating and suspended debris. Escape from natural rivers can be difficult enough, but the shape, size, water speeds, and smooth concrete lining of flood channels frequently make escape impossible without physical rescue. Because many flood channels flow through highly populated areas, people are washed away every year. These rescues are, without a doubt, among the most dangerous situations firefighters and other rescuers will ever confront.

Multi-tier response

In the LACoFD system, first responders are reinforced by a *second tier*, which includes two USAR task force stations, a fleet of seven Bell 412 helicopters, two Firehawks, and a Long Ranger. There are also up to eighteen specially trained airborne and ground-based, swift-water rescue teams based at strategically chosen fire stations across the county.

When combined with a county-wide network of existing fire department-based USAR units, mountain SAR teams from the county Sheriff's Department and the unique capabilities of various public works agencies ensure that any swift-water rescue or flood incident in L.A. County will be met with a timely and effective emergency response.

The flood-control channel problem

In the early 1980s, a number of swift-water rescue instructors from other regions were stumped when asked by LACoFD firefighters:

"What are we to do about the problem of children being swept away by water moving 30 to 40 miles per hour in vertical-sided, concrete-lined flood-control channels?" Frustrated by the lack of training, equipment, and rescue procedures for flood channel operations, water rescue committee members embarked on a years-long mission of research, development, and testing to establish new methods to rescue people from flood-control channels.

Flood-control channels are one of the most dangerous swift-water rescue hazards in southern California and the other cities served and protected by these systems. Flood-control channels are essentially man-made or natural waterways that have been lined in concrete, rock, or other materials. They are designed to drain the greatest volume of water to the sea in the quickest possible manner. In other words, they are engineered to be high-velocity, high-volume waterways that will carry more water than natural waterways of similar size and length.

In most instances, flood-control channels are built without regard for anyone who might become trapped in their grip; in part because it is illegal for people to stray into them and because they weren't designed for human recreation. Despite this legal obstacle, children and adults in many areas of North America use flood channels as short cuts to school and work, as bike paths and running paths, and as recreation areas. And every year, dozens of people are swept away in flood channels in places like L.A. and Albuquerque.

Few natural rivers in the world can match the velocity, water volume, and sheer killing power of concrete-lined, flood-control channels, especially those that drain water from high mountains to the oceans or deserts. Even the waterway rescue preplans developed in the Midwest did not take into account the unique dangers and conditions created by water moving through vertical-sided, concrete-lined channels with elevation drops of thousands of feet in a distance of 50 miles,[7] channels capable of flowing more water than the Mississippi River at freeways speeds.

Even in southern California, a place known for its attention to emergency preparedness, the unique challenges presented by flood-control channels had previously not been addressed in any meaningful detail. This was caused in part because formal responsibility for these flood channel rescues was not placed specifically on any particular agency and also because there were few people available with the experience, training, and knowledge to come up with effective solutions.

Los Angeles/New Mexico collaboration bears fruit

The firefighters eventually found solutions to most of the flood-control channel problems they encountered, and some of those solutions came from a unique collaboration with firefighters from Albuquerque, New Mexico. For decades Albuquerque firefighters have confronted flood-control channel rescue problems similar to those faced by their counterparts in southern California. Both places are home to densely populated urban areas that grew up in the shadows of steep mountain ranges that tower over their skylines. In both cases, the local mountains shed immense volumes of water toward the populated areas whenever significant rains occurred. Both places are subject to intense storms that cause catastrophic flash floods.

Even when it's sunny in Albuquerque or L.A., thunderstorms in the mountains can send walls of water rushing down the canyons into the populated areas. In L.A. County, the lethal effects of flash flooding are tragically

demonstrated by incidents like the 7-ft high wall of water that swept five children down the Alhambra Wash in November 1997, leaving three dead and trapping two firefighters on an island.[8] In both Albuquerque and southern California, flood control is based on the development and maintenance of complex systems of concrete-lined channels designed to divert runoff from the cities and funnel the rainwater to the ocean or major rivers in the quickest manner possible.

In New Mexico, the steep gradient of the land between the mountains and the Rio Grande River causes storm water and snow melt to cascade through flood-control channels in Albuquerque at velocities that sometimes exceed 30 or 40 miles per hour. In southern California, the same phenomenon causes water to funnel between the 10,000-ft San Gabriel Mountains and the Pacific Ocean in a distance of just over 50 miles. The resulting water velocities are deadly, sometimes exerting thousands of pounds of force against any person who becomes trapped against the upstream side of a rock, tree, or any other stationary obstacle. Striking such an object in these currents often proves fatal, and anyone who survives the impact usually finds that escape is impossible without assistance from specially trained swift-water rescue teams.

During the course of attempting rescue in flood channels, firefighters both in Albuquerque and L.A. County have suffered a significant number of near-fatal accidents when they jumped in to rescue victims (or accidentally fell into the water during the course of shore-based rescue attempts).

Such was the case on February 6, 1998, when an Ontario (California) firefighter fell into a flood channel while attempting to rescue a man and his dog. The firefighter traveled more than 7 miles with the original victim before both were rescued; the victim was ultimately swept nearly 12 miles. The Ontario rescue is rather typical. In fact, it is not at all uncommon for flood channel rescue operations to extend for between 15 and 30 miles and require the response of dozens of fire and rescue units. In both L.A. County and Albuquerque, firefighters have demonstrated that these incidents are survivable for many victims who are the beneficiaries of quick, effective rescue response.

Albuquerque firefighters spent years developing a city-wide rescue system that relies on the use of innovative waterway rescue preplans and the pioneering use of new equipment and techniques. Ironically, the LACoFD Water Rescue Committee, which had been working to develop some sort of preplan system, discovered in 1986 that many of the developments in Albuquerque paralleled those in L.A. County. Phone calls, information, and visits were exchanged, and what ensued was a successful cooperative program by which both agencies shared information, experiences, and results from innovative approaches to common problems.

One of the most important products of this partnership was development of the water rescue curtain that is now used across L.A. County to snatch victims from the roiling waters. The first curtain was developed and built by Albuquerque Fire Captain (now Assistant Chief) Ted Nee, who stitched the thing together in his own garage after observing a failed rescue in his city.

Nee's idea was to equip strategically-located engine companies with the rescue curtain to be quickly set up by firefighters responding to locations downstream of victims being swept away in flood-control channels to establish

rescue points. The first action in this system is to pass a rescue rope across the channel using line-throwing guns, helicopters, or a bridge to traverse the channel. The rope is quickly anchored to sturdy objects to create a 60% (or greater) angle in relation to the current.

Then firefighters establish a mechanical advantage system to tension the rope until it reaches a sufficient level of tautness. Through the use of caribiners used as hangers, the rescue curtain is suspended from the tensioned rope so that its lower section dangles at the surface of the water. When a victim passes a rescue curtain site, he is likely to become entangled in the rectangular-shaped mesh, which quickly zips him to shore, powered by the water pushing on the victim and the curtain together. The effect is not unlike that of a shower curtain sliding across a horizontal shower rod.

Nee enlisted the help of scientists from Sandia National Testing Laboratories to develop computer models of the forces placed on the rescue curtain when it is positioned at different angles to the current to ascertain the optimum working angles and other factors. After months of discussion and idea exchanges with this author, the Albuquerque Fire Department flew Nee to L.A. in 1985 to field-test the new system with members of the LACoFD's Water Rescue Committee. The result of this testing and modeling was a system that has proven to be effective for removing victims from fast-moving water. In fact, it was found that higher velocities actually propel the curtain (and victims) to shore more efficiently.

As a result of the computer modeling and field-testing, additional tag line ropes were added to give firefighters the option of pulling the lines to increase the speed with which the victim is transported to shore. Now, when the victim and the curtain reach the shoreline, specially equipped firefighters are there to pull them from the water. If necessary, the rescue curtain can be quickly repositioned over the water to rescue additional victims.

Today, rescue curtains are strategically located on fire engines and ladder trucks across L.A. County. Their use has proven effective not only to rescue victims but to provide downstream safety for rescuers engaged in hazardous rescue operations. This is one example of addressing the unique problems created by flood-control channels.

Waterway rescue pre-plans

An important element of LACoFD's swift-water system is its use of waterway rescue preplans that have been developed for virtually every major flood-control channel and river in the department's jurisdictional area, something never been done before in southern California. The preplans are designed to give firefighters and rescue teams a sort of *blueprint* for the immediate deployment of resources to search for victims and simultaneously establish strategically-located rescue points along the course of the involved waterway.

Waterway rescue preplans show how to best access channels and the location of life-threatening hazards, including strainers and low head dams. Preplans also include the average maximum speed of the water with attendant timetables that enable rescuers to anticipate how long it will take a victim to reach various points in the river. It can help identify pre-designated locations at which helicopter-based swift-water rescue teams can swoop down to pluck victims from the current without becoming entangled in power lines and other hazards.

One of the most innovative strategies to spring from the LACoFD's collaborations with other agencies was the original concept for waterway rescue preplans. The Ohio Department of Natural Resources was one of the early leaders in the drive to develop written plans to provide a sort of blueprint for SAR operations along natural waterways. The intent was to make the best use of local knowledge about rivers and emergency resources to ensure the quickest, most effective response to river rescue incidents. The LACoFD adopted this approach to flood-control channels and other urban and rural waterways.

Typically, waterway rescue preplans include clearly identified rescue points, or places known to provide rescuers with an advantage over the river. Preplans can identify good anchor points for rope systems, access points for bridges over the target channels, and other characteristics that can be used to the rescuer's advantage. They also include clearly defined danger points where rescuers might be killed or injured if they stray there. One of the committee's main challenges—and ultimately among its chief contributions—was the development of waterway rescue preplans for all of the major waterways in L.A. County.

First responder training

In 1984, LACoFD administrators approved a plan proposed by the department's water rescue committee to develop a series of in-house training courses to teach first responder firefighters to conduct shore-based rescue operations. The plan included the development of written preplans for rescues along the county's 500 miles of flood channels and a way to convey the most important lessons learned by two years of intense research and development.

At first, this training was received with a certain level of skepticism by some of the firefighters who were schooled in a tradition that emphasized firefighting, vehicle extrication, and emergency medical services. For some of them, it seemed unfeasible for the fire department to take on the additional responsibility for managing water rescue incidents. Some argued that there were already too many responsibilities placed on the fire department. In their view, the swift-water rescue program would further dilute the ability of firefighters to deal with fires and other *traditional* fire department mission objectives.

The counterpoint, of course, was that whenever the public dials 911, the fire department always responds without delay and without regard to the type of rescue or fire being reported. Because fire stations are strategically located to provide the most timely response to any point within most populated areas, fire department units are a natural choice to assume primary responsibility for managing swift-water rescue incidents. The assumption of responsibility for swift-water rescue was a natural outgrowth of the fire department's rapidly expanding role as the primary provider of technical rescue and disaster rescue service.

Hundreds of daily technical rescues and dozens of disaster-related rescue operations have been properly managed by local fire departments in recent years. The lead role played by fire departments in technical rescue has proved to be optimum in terms of timeliness of emergency response, the ability to mobilize massive resources for disasters, the cost effectiveness of tax dollars devoted to dual-function fire/rescue units, and the ability of rescue-trained firefighters to pull live victims from horrendous rescue entrapment situations that might have been deemed hopeless in past years.

Thus, much of the impetus for current movement toward fire department-based technical rescue capabilities found much of its momentum in the early efforts of fire departments to establish formal swift-water rescue, technical rope rescue, and building collapse rescue programs in the early 1980s.

Locally, these capabilities have been further enhanced since 1987, when LACoFD and a number of other fire departments in L.A. County adopted comprehensive USAR programs to effectively manage the entire range of technical rescues that occur there. These rescues include earthquakes, mudslides, transportation accidents, high-angle rescues, construction accidents, and swift-water rescue incidents. In particular, the USAR units provide a second, more advanced tier of response to swift-water and flood rescue operations wherever they may occur in L.A. County.

Appendix I: California USAR/Rescue Resource-typing Levels[9]

Introduction

The USAR organizational module is designed to provide supervision and control of essential functions at incidents where technical rescue expertise and equipment are required for safe and effective rescue operations. USAR incidents can be caused by a variety of events such as earthquakes, floods, and hurricanes that cause widespread damage to a variety of structures and entrap hundreds of people. Other examples of USAR incidents can range from mass transportation accidents with multiple victims to single-site events such as trench cave-in and confined-space rescue operations involving only a few victims. USAR operations are unique in that specialized training and equipment are required to mitigate the incident in the safest and most efficient manner possible.

Initial USAR operations will be directed by the first arriving company officer who will assume command as the IC. Subsequent changes in the incident command structure will be based on the resource and management needs of the incident following established ICS procedures.

Additional resources may include USAR companies and USAR crews specifically trained and equipped for USAR operations. The USAR company is capable of conducting SAR operations at incidents where technical expertise and equipment are required. USAR crews are trained USAR personnel dispatched to the incident without rescue equipment. USAR companies and crews can be assigned as a single increment, grouped to form USAR strike teams, or added to other resources to form a task force. USAR single increments, strike teams, and task forces are managed the same as other incident resources.

Due to the unique hazards and complexity of USAR incidents, the IC may need to request a wide variety and amount of multi-disciplinary resources.

USAR companies and crews have been categorized or *typed*. Typing is based on an identified operational capability. Four levels of USAR operational capability have been identified to assist the IC in requesting appropriate resources for the incident. These levels are based on four general construction categories and related incidents the rescuers may encounter, and identifies minimum training and equipment required for safe and effective rescue operations in these situations. Levels of USAR operational capability and general construction categories are identified in the Glossary of Terms and attachments A and B.

USAR incidents may occur that will require rescue operations that exceed a resource's identified capability. When the magnitude or type of incident is not commensurate with a capability level, the IC will have the flexibility to conduct rescue operations in a safe and appropriate manner using existing resources within the scope of their training and equipment until adequate resources can be obtained or the incident is terminated.

Unified command

A unified command structure may need to be utilized at USAR incidents due to the involvement of multiple agencies and jurisdictions having statutory or political responsibility or authority. A unified command located at a single command post is the best method (February, 1995 ICS-USAR-120-12) for ensuring effective information flow, coordination, safety, and maximum utilization of resources, which can reduce fiscal impact.

Example:

An example of a USAR incident involving a multi-agency and multi-jurisdictional response is an earthquake causing the collapse and damage of several structures over a large but confined area that crosses jurisdictional boundaries. The event has trapped multiple victims in densely populated areas of the city, which contracts its law enforcement capability from the adjacent county. In this city, the fire department will have responsibility for fire suppression, initial medical treatment, and SAR. Law enforcement will be responsible for scene security, traffic control, and evacuation. Additional resources from a variety of agencies and organizations will be required to mitigate the incident. A unified command structure will ensure effective coordination and utilization of each responding resource.

ICS modular development

The flexibility and modular expansion capabilities of the ICS provide an almost infinite number of ways USAR resources can be arranged and managed. A series of modular development examples are included to illustrate one possible method of expanding the incident organization based on the example scenario described above.

The ICS modular development examples shown are not meant to be restrictive, nor imply these are the only ways to build an ICS organizational structure to manage USAR resources at an incident. To the contrary, the ICS modular development examples are provided only to show conceptually how one can arrange and manage resources at a USAR incident that builds from an initial response to a multi-branch organization.

ICS modular development examples

- *Initial response organization*—The first-to-arrive fire department company officer will assume command of the incident as the IC. Initial-response resources are managed by the IC who will assume all command and general staff functions and responsibilities.

• *Reinforced response organization*—In addition to the initial response, more law enforcement, local engine and truck companies, and mutual aid resources have arrived. The IC has established a safety officer to assure personnel safety and a public information officer to manage the large media presence. A staging area is established to check-in arriving resources. The incident is geographically divided into Divisions A and B to better manage resources. The original engine and truck companies are grouped together to form Task Force #1. Second to arrive local engine and truck companies are grouped together to form Task Force #2. Public works is removing debris from the street to improve access and egress routes. Examples of possible assigned functions are enclosed in brackets below each resource.

• *Multi-group/division response organization*—The IC forms a unified command with the senior ranking law enforcement official on scene, has added a liaison officer to the command staff to coordinate assisting agencies' participation, and assigned an operations and planning section chief. Several operational units have been formed to better coordinate the large amount of resources at the incident.

A law group and medical group have been formed. A structural engineer technical specialist is assisting Division B resources with structural damage assessment. A hand crew strike team is conducting debris removal. One state/national USAR task force has arrived and is assigned to Division A. One USAR technical specialist who understands the unique complexities and resource requirements at USAR incidents has been assigned to the planning section.

• *Multi-branch response organization*—The IC has assigned a logistics and finance/administrative section chief. The operations section has established five branches with similar functions to

better coordinate and manage resources. The planning, logistics and finance/administrative sections have several units operational to support the large number of resources at the incident.

Attachment A of Appendix I:

Four General Types of Building Construction

The construction types and occupancy usage of various structures may require the use of a variety of different techniques and materials. The four general construction categories the rescuer will most likely encounter in collapse situations are light frame, heavy wall, heavy floor and pre-cast concrete construction. These four general classifications of construction usually comprise the majority of structures affected by collapse and failure.

Light frame construction

Materials used for construction are generally lightweight and provide a high degree of structural flexibility to applied forces such as earthquakes, hurricanes, tornadoes, etc. These structures are typically constructed with a skeletal structural frame system of wood or light-gauge steel components that provide support to the floor or roof assemblies. Examples of this construction type are wood frame structures used for residential, multiple low-rise occupancies and light commercial occupancies up to four stories in height. Light-gauge steel frame buildings include commercial business and light manufacturing occupancies and facilities.

Heavy wall construction

Materials used for construction are generally heavy and use an interdependent structural or monolithic system. These types of materials and their assemblies tend to make the structural system inherently rigid. This construction type is usually built without a skeletal structural frame. It uses a heavy wall support and assembly system to provide support for the floors and roof assemblies. Occupancies using tilt-up concrete construction are typically one to three stories in height and consist of multiple monolithic concrete wall panel assemblies. They also use an interdependent girder, column, and beam system for providing lateral wall support of floor and roof assemblies. Occupancies typically include commercial, mercantile, and industrial.

Other examples of this type of construction include reinforced and URM buildings, typically of low-rise construction, one to six stories in height, of any type occupancy.

Heavy floor construction

Structures of this type are built using cast-in-place concrete construction consisting of flat slab panel, waffle, or two-way concrete slab assemblies. Pre-tensioned or post-tensioned reinforcing steel rebar or cable systems are common components for structural integrity.

The vertical structural supports include integrated concrete columns and concrete enclosed or steel frames that carry the load of all floor and roof assemblies. This type includes heavy timber construction that may use steel rods for reinforcing. Examples of this type of construction include offices, schools, apartments, hospitals, parking structures, and multipurpose facilities. Common heights vary from single story to high-rise structures.

Pre–cast construction

Structures of this type are built using modular pre-cast concrete components that include floors, walls, columns,

and other sub-components that are field-connected upon placement on site. Individual concrete components use embedded steel reinforcing rods and welded wire mesh for structural integrity and may have either steel beam, column, or concrete framing systems used for the overall structural assembly and building enclosure. These structures rely on single- or multi-point connections for floor and wall enclosure assembly and are a safety and operational concern during collapse operations. Examples of this type of construction include commercial, mercantile, office, and multi-use or multi-function structures, including parking structures and large occupancy facilities.

Attachment B of Appendix I:

Four Levels of USAR Operational Capability

Basic operational level

The basic level represents the minimum capability to conduct safe and effective SAR operations at structure collapse incidents. Personnel at this level shall be competent at surface rescue that involves minimal removal of debris and building contents to extricate easily accessible victims from non-collapsed structures.

Light operational level

The light level represents the minimum capability to conduct safe and effective SAR operations at structure collapse incidents involving the collapse or failure of light frame construction and basic rope rescue operations.

Medium operational level

The medium level represents the minimum capability to conduct safe and effective SAR operations at structure collapse incidents involving the collapse or failure of reinforced and URM, concrete tilt-up, and heavy timber construction.

Heavy operational level

The heavy level represents the minimum capability to conduct safe and effective SAR operations at structure collapse incidents involving the collapse or failure of reinforced concrete or steel frame construction and confined-space rescue operations.

Attachment C of Appendix I:

Four Levels of USAR Operational Capability Minimum Training

Basic operational level

The basic operational level represents the minimum capability to operate safely and effectively at a structural collapse incident. Personnel at this level shall be competent at surface rescue and rescue involving minimal removal of debris and building contents to extricate easily accessible victims from non-collapsed structures. Rescue operations would include removal of victims from under furniture, appliances, and the surface of a debris pile.

Training at the basic level should at a minimum include the following:

A. Size-up of existing and potential conditions and the identification of the resources necessary to conduct safe and effective USAR operations.

B. Process for implementing the ICS.

C. Procedures for the acquisition, coordination, and utilization of resources.

D. Procedures for implementing site control and scene management.

E. Identification, use, and proper care of PPE required for operations at structural collapse incidents.

F. Identification of construction types and the characteristics and expected behavior of each type in a collapse incident.

G. Identification of four types of collapse patterns and potential victim locations.

H. Recognition of the potential for secondary collapse.

I. Recognition of the general hazards associated with a structural collapse and the actions necessary for the safe mitigation of those hazards.

J. Procedures for implementation of a structural identification marking system and a structural hazard marking system.

K. Procedures for conducting searches at structural collapse incidents using appropriate methods for the type of collapse.

L. Procedures for implementation of a search marking system.

M. Procedures for the extrication of victims from structural collapse incidents.

N. Procedures for providing initial medical care to victims.

Light operational level

Personnel shall meet all basic level training requirements. In addition, personnel shall be trained in hazard recognition, equipment use, and techniques required to operate safely and effectively at structural collapse incidents involving the collapse or failure of light frame construction and basic rope rescue as specified below:

A. Personnel shall be trained to recognize the unique hazards associated with the collapse or failure of light frame construction. Training should include but not be limited to the following:

- Recognition of the building materials and structural components associated with light frame construction

- Recognition of unstable collapse and failure zones of light frame ordinary construction

- Recognition of collapse patterns and probable victim locations associated with light frame construction

B. Personnel shall have a working knowledge of the resources and procedures for performing search operations intended to locate victims who are not readily visible and who are trapped inside and beneath debris of light frame construction. Training should include but not be limited to the following:

- Types of search resources: USAR dogs, optical instruments (search cameras), seismic/acoustic instruments (listening devices)

- Capabilities of search resources

- Acquisition of search resources

C. Personnel shall be trained in the procedures for performing access operations intended to reach victims trapped inside and beneath debris associated with light frame construction. Training should include but not be limited to the following:

- Lifting techniques to safely and efficiently lift structural components of walls, floors, or roofs

- Shoring techniques to safely and efficiently construct temporary structures needed to stabilize and support structural components to prevent movement of walls, floors, or roofs

- Breaching techniques to safely and efficiently create openings in structural components of walls, floors, or roofs

- Operating appropriate tools and equipment to safely and efficiently accomplish the above tasks.

D. Personnel shall be trained in the procedures for performing extrication operations involving packaging, treating, and removing victims trapped inside and beneath debris associated with light frame construction. Training should include but not be limited to the following:

- Packaging victims within confined areas

- Removing victims from elevated or below grade areas

- Providing initial medical treatment to victims at a minimum to the basic life support (BLS) level

- Operating appropriate tools and equipment to safely and efficiently accomplish the above tasks

Medium operational level

Personnel shall meet all light level training requirements. In addition, personnel shall be trained in hazard recognition, equipment use, and techniques required to operate safely and effectively at structural collapse incidents involving the collapse or failure of reinforced and URM, concrete tilt-up, and heavy timber construction.

Heavy operational level

Personnel shall meet all medium level training requirements. In addition, personnel shall be trained in hazard recognition, equipment use, and techniques required to operate safely and effectively at structural collapse incidents involving the collapse or failure of reinforced concrete or steel frame construction and confined-space rescue.

Endnotes

1 www.firescope.oes.ca.us.gov Incident Command System-Urban Search and Rescue Operational System Description (ICS-USAR-120-1). OES FIRESCOPE OCC Document Control Unit, 2524 Mulberry Street, Riverside, California, 92501-2200 (909) 782-4174 Fax (909) 782-4239.

2 The LACoFD fields one of eight California-based FEMA USAR task forces (part of the network of 28 USAR task forces for response to domestic disasters). The LACoFD USAR task force is one of the U.S. State Department (Agency for International Development, Office of Foreign Disaster Assistance) teams for international disaster response.

3 A similar event occurring today, with nearly 100,000 people living in the St. Francis Dam's original inundation area would be even more devastating.

4 *Type 1 (Heavy) USAR company* is a standard California term that identifies six-member units equipped and trained to perform SAR in collapsed reinforced concrete and steel structures, confined spaces, and other complicated technical rescue situations

5 Czajkowski, John. 2001. "NFPA 1670 Hits the Streets." *Fire Rescue Magazine* (March).

6 Sargent, Chase. 1999. "NFPA 1670, New Standards For Technical Rescue." *Fire Engineering Magazine* (October).

7 To illustrate this point, consider the following fact: within a distance of only 50 miles, the Los Angeles River, which flows from the San Gabriel Mountains through downtown L.A. before emptying into the Pacific in Long Beach, encounters an elevation drop exceeding the elevation drop measured along the entire length of the Mississippi River.

8 A helicopter-based LACoFD swift-water rescue team rescued both firefighters.

9 California Governor's OES. FIRESCOPE ICS USAR 120-1 (Appendix B); ICS 420-1; and USAR O.E.S. Resource Evaluation Form.

4

Rescue Apparatus and Equipment

If there's one area in which rescue is progressing at the proverbial speed of light, it's the arena of apparatus and equipment. Today there is such a wide variety of choices in the design and construction of rescue apparatus, and there are so many options for rescue equipment, that fire/rescue agencies are well advised to consider their needs far into the future before making expensive selections (Fig. 4–1).

Fig. 4–1 The Evolution of L.A. USAR Apparatus, 1990–2002

Some fire/rescue agencies may find it advantageous to make large purchases (especially those involving emerging technology and expensive rescue and USAR

apparatus) in increments to reduce annual costs. This also allows the purchase of upgraded or different models as SAR technologies improve and as manufacturers recognize that building better rescue tools for an increasingly receptive fire/rescue community simply makes good business sense. Items that seem to fit your needs now may be outdated next year, and antiquated just a few years hence, so some research is in order to make sure the best apparatus and equipment are selected (Fig. 4–2).

Fig. 4–2 Remote Search Camera

Manufacturers have come to realize that the field of rescue apparatus and equipment is profitable because the demand is increasing as more fire departments and state governments come online with formal rescue or USAR programs. In addition, FEMA and other federal agencies continue to improve the nation's urban SAR capabilities to deal with the consequences of natural and man-made disasters (especially the emerging consequences of terrorism). As more scientists and researchers put their minds to improving rescue, new devices are being created to help us do our jobs more effectively. And as additional military-grade research and technology is released for use by public safety agencies, it is being used by manufacturers to develop better tools for rescue (Fig. 4–3).

Fig. 4–4 In Some Areas, Helicopters are Used to Conduct Technical Rescues on a Daily Basis

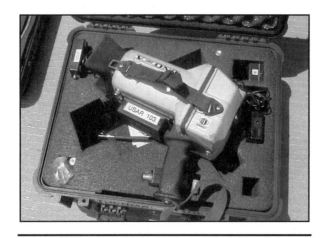

Fig. 4–3 Thermal Imaging Systems are Now a Standard Item on USAR Units and Rescue Companies

Some obvious examples of *transfer of technology* from military to civilian public safety use include thermal imaging, night vision, global positioning satellite (GPS) technology, new fabrics and materials, chemical/biological agent detection, hand-held computing devices, military helicopters like the Blackhawk (developed for fire/rescue as the Firehawk),[1] and better communications systems like satellite phones and the like (Fig. 4–4).

Some not-so-obvious emerging technologies adapted from military use are likely to include highly sensitive and portable ground-penetrating radar for locating trapped victims, virtual-reality-based rescue training, and better aerial surveillance. Highly advanced personnel accountability systems will monitor the exact location (and even the vital signs) of firefighters and rescuers and *bread crumb* systems will help lead lost firefighters and rescuers back to safety. There will also be advances in lightweight materials for PPE and other innovations still on the drawing board.

Firefighters are increasingly vocal about demanding better tools with which to do the job of rescue, and designers and manufacturers are stepping up to the plate. Today the field of rescue apparatus and equipment is loaded with independent manufacturers providing new and innovative tools. Companies manufacturing mountaineering, automotive, geographic information systems, imaging, communications, chemical/biological sensing, and extreme recreation equipment are also offering improved rescue and firefighting products that would have seemed farcical just a few decades ago (Fig. 4–5).

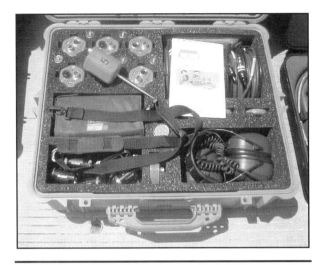

Fig. 4–5 Sensors Used to Locate Victims Trapped in Collapses

Because of the emergence of vast new technologies combined with a new emphasis on better apparatus and equipment, the intent of this chapter is to provide an overview of some current and future options. This chapter should be read with the understanding that the items shown and discussed may very well be outdated in just a few years. This new outpouring of ever-improving products is a hopeful sign because it means that designers and manufacturers are listening to *you* and providing the tools, apparatus, and equipment that you will need to do your job better and with a wider margin of safety. If all the items shown in this chapter were still state of the art two or three years hence, it would be a problem (Fig. 4–6).

Fig. 4–6 Modern USAR Apparatus Carry a Multitude of Tools for Cutting, Breaching, Lifting, and Stabilization

Rescue/USAR Apparatus

The choice of rescue and USAR apparatus must be based on factors such as: local terrain and geography (road conditions), local weather, anticipated personnel requirements (including local, regional, state, or federal guidelines), current and future equipment needs (including those specified by local, regional, state, or federal guidelines), safety requirements like NFPA 1500 (Firefighter Safety and Survival), and of course, budget. There is also the issue of past practice, or tradition, which sometimes dictates choices. Unfortunately, budget concerns sometimes override some of the other issues and compromises may be necessary.

There are several broad categories of rescue apparatus, which we can call straight body, walk-through, partial walk-through, non-walk-through, P.O.D., tillered, and fifth-wheel or tractor-trailer. There are a wide variety of commercial vehicles and fire apparatus adapted to rescue, and there are others that we can call specialized (Fig. 4–7).

Fig. 4–7 An Example of the Classic Modern Rescue Company

Fig. 4–8 This Unit is of the "Straight Body Walk-through" Design Favored by FDNY

Rescue Apparatus

Straight body rescue apparatus— walk-through

Advantages. Walk-through, straight body rescue apparatus (Fig. 4–8) offer the following advantages:

- Protection from the elements and privacy while personnel are donning or changing PPE

- Face-to-face communication during transit to and from emergency incidents, training, etc.

- Protection of sensitive equipment from the elements

- A place for crew briefings, planning, strategy sessions

- Straight body design is advantageous on rough terrain, on motorways, in floods, on earthquake-damaged or debris-strewn roads, etc.

- Potential for 4-wheel drive (or all-wheel drive) design

- In use throughout the world with proven results

Disadvantages. Walk-through, straight body rescue apparatus have the following disadvantages:

- Reduced overall space for equipment storage due to space devoted to walk-in.

- Possible personnel safety issues in the event of a crash.

- Not as maneuverable as tillered rescue apparatus in tight spaces, narrow roads, etc.

Straight body rescue apparatus— partial walk-through

Advantages. Partial walk-through, straight body rescue apparatus (Fig. 4–9) offer the following advantages:

- Protection from the elements and privacy while personnel are donning or changing PPE

- Additional overall storage space for equipment compared with full walk-in apparatus

- Face-to-face communication during transit to and from emergency incidents, training, etc.

- Protection of sensitive equipment from the elements

- A place for crew briefings, planning, strategy sessions

- Straight body design is advantageous on rough terrain, on motorways, in floods, on earthquake-damaged or debris-strewn roads, etc.

- Potential for 4-wheel drive (or all-wheel drive) design

- In use throughout the world with proven results

Straight body rescue apparatus—non–walk-through

Advantages. Non-walk-through, straight body rescue apparatus (Fig. 4–10) offer the following advantages:

- Maximum overall area for equipment storage

- Straight body design is advantageous on rough terrain, on motorways, in floods, on earthquake-damaged or debris-strewn roads, etc.

- Potential for 4-wheel drive (or all-wheel drive) design

- In use throughout the world with proven results

Fig. 4–9 This "Partial Walk-through Straight Body" Apparatus Has a Shorter Wheelbase for Maneuverability

Fig. 4–10 The Sacramento Metro FD Favors a Longer Wheelbase with No Walk-through for More Equipment Storage

Disadvantages. Partial walk-through, straight body rescue apparatus have the following disadvantages:

- Reduced overall space for equipment storage due to space devoted to partial walk-in.

- Reduced area for donning and changing PPE.

- Not as maneuverable as tillered rescue apparatus in tight spaces, narrow roads, etc.

Disadvantages. Non-walk-through, straight body rescue apparatus have the following disadvantages:

- Less protection for sensitive equipment that would normally be stored in compartments accessible from the inside walk-through area

- Less protection from the elements and privacy while donning and changing personal protective equipment

- Not as maneuverable as tillered rescue apparatus in tight spaces, narrow roads, etc.

POD rescue apparatus

Advantages. POD rescue apparatus offer the following advantages:

- May be advantageous on rough terrain, on motorways, in floods, on earthquake-damaged or debris-strewn roads, etc.

- Potential for 4-wheel drive (or all-wheel drive) design

- In use throughout the world with proven results

Disadvantages. POD rescue apparatus have the following disadvantages:

- Limited selection of equipment when it arrives on the scene, particularly if the incident is a multi-hazard one, or if it has been categorized in such a way by dispatchers that the wrong POD is selected.

- Possible error in selecting the proper POD based on information from the reporting party and interpretation by the dispatchers.

- Time required to select, load, deliver, and off-load the POD.

- Possible malfunction of POD loading/offloading mechanism can delay operations.

Tillered rescue apparatus

Advantages. Tillered rescue apparatus offer the following advantages:

- Far more maneuverable in tight spaces, on narrow roads, etc.

- Huge amounts of potential storage space for equipment

- Tillered design may be advantageous on rough terrain, on motorways, in floods, on earthquake-damaged or debris-strewn roads, etc.

- Tillered fire aerial ladder apparatus is in use around the world with proven results

- Requires a minimum of three personnel (engineer or chauffeur, tiller driver, officer)

Disadvantages. Tillered rescue apparatus have the following disadvantages:

- New concept for rescue apparatus that is not yet in use around the world with proven results

- Relatively few manufacturers have built tillered rescue apparatus

- More costly than some other designs

- Requires a minimum of three personnel (engineer or chauffeur, tiller driver, officer)

Fifth-wheel or tractor-trailer rescue apparatus

Advantages. Fifth-wheel or tractor-trailer rescue apparatus offer the following advantages:

- Ability to swap tractors (if additional tractors have been purchased or are available). During maintenance and repairs, keeping the trailer carrying the bulk of the rescue equipment in service without the need to switch the equipment to a reserve apparatus.

- Ability to swap tractors to continue response in case of mechanical breakdown of tractor

- Cost savings in comparison with some other designs

- In use in many parts of the United States with mixed results

In Figure 4–11, a fifth-wheel USAR apparatus is equipped with dual capstan systems, one on each side of the rig, each driven by a power take-off system and controlled by a firefighter operating a *dead mans switch*-type joystick. It is also equipped with a rescue winch system for vehicles and larger loads.

Fig. 4–11 This "Straight-body without Walk-through" Apparatus is Outfitted with Anchor Points for High-angle Rescue and a Motor Mount for Its Inflatable Rescue Boat

Fig. 4–12 illustrates how multiple anchor points are important for high-angle rescue operations. This USAR apparatus is set up for operation of the dual capstan systems, each with a separate belay/safety line anchored to the bumper. It is also set up for operation of a rescue winch system.

Fig. 4–12 Fifth-Wheel USAR Apparatus for Increased Maneuverability

Disadvantages. Fifth-wheel or tractor-trailer rescue apparatus have the following disadvantages:

- In use in many parts of the United States with mixed results

- Disadvantageous on rough terrain, on motorways, in floods, on earthquake-damaged or debris-strewn roads, etc.

Commercial Vehicles and Fire Apparatus Adapted for Rescue

The Long Beach (CA) Fire Department designed an engine company for technical rescue operations based on FDNY squad company specifications. The new company adapted a commercial truck for use as a USAR vehicle (Fig. 4–13) to create a USAR task force.

Advantages. Adapting a commercial truck for use as a rescue apparatus offers the following advantages:

- Cost savings by dual-function apparatus

- May be advantageous on rough terrain, on motorways, in floods, on earthquake-damaged or debris-strewn roads, etc.

- Potential for 4-wheel drive (or all-wheel drive) design

Disadvantages. Adapting a commercial truck for use as a rescue apparatus may have the following disadvantages:

- May not suit the rescue mission completely because of dual-function design

- May lack adequate storage space for rescue equipment

- May not have proven track record for rescue use

Fig. 4–13 This Fifth-wheel Design Allows for Immense Equipment-carrying Capacity

Specialized rescue apparatus

Advantages. Specialized rescue apparatus (Fig. 4–14) offer the following advantages:

- Special or unique design to meet the demands of specific conditions and hazards

- May be advantageous on rough terrain, on motorways, in floods, on earthquake-damaged or debris-strewn roads, etc.

- Potential for 4-wheel drive (or all-wheel drive) design

Disadvantages. Specialized rescue apparatus may have the following disadvantages:

- Potentially more costly

- Similar or like models may not be widely used and may not have proven track record

Fig. 4–14 This Capstan and Rescue Winch System is an Example of the Specialty Items that May Be Incorporated in Modern Rescue Apparatus

Combinations of rescue/USAR apparatus

In cases where all the equipment cannot be carried on the rescue or USAR company apparatus, some agencies assign support vehicles and trailers to carry additional or specialized equipment such as IRBs, PWC, timbers for cribbing and shoring, etc. (Fig. 4–15).

Fig. 4–15 Minimum Equipment Complement Required for California Type I USAR Companies (Courtesy Jim Hone).

Some agencies, including the LACoFD, combine one USAR-trained engine company *and* one quick-response 4-wheel drive USAR unit as USAR task force fire stations. This concept ensures timely response of the USAR company captain and USAR task force firefighters to distant incidents to provide immediate technical assistance and to get the ball rolling on complex technical rescue operations. Meanwhile, the USAR company and USAR engineer respond in their larger apparatus (Fig. 3–16).

Fig. 4–16 An Example of a Combination Apparatus

This system ensures the ability of USAR task force fire station personnel to get to the scene of rescues in rugged mountainous areas that may be flooded and require 4-wheel drive vehicles or have roads too narrow to accommodate heavy rescue apparatus. At the scene of multi-alarm fires, at least one LACoFD USAR task force is always assigned (usually) to augment the RIC and help ensure firefighter safety and rescue. The smaller USAR unit may be maneuvered around hose lines and larger apparatus to get into choice position for RIC and other assigned operations.

Thus, LACoFD USAR task forces often respond in convoys of three vehicles. It may seem somewhat awkward at first, but this approach has proven to give the USAR captains more flexibility in getting to the scene in the most timely manner. Their other option is flying in department fire/rescue helicopters to manage complex and/or long-term emergency operations (Fig. 4–17).

Fig. 4–17 An Example of Customized All-terrain Vehicles for Special Hazards and Terrain

Other specialized vehicles for rescue may include bulldozers, track hoes, cranes, compressor trailers, hydro-vac trucks, shoring trailers, and practically any other vehicle used for construction, demolition, road building, or mining, etc. A specialized apparatus used for rescue is the multi-processor.

Rescue Equipment

In some states that experience a high volume of technical rescues and rescue-related disasters, efforts have been made to standardize the minimum equipment complements for various types of USAR or rescue units. Some states have convened working groups or committees to investigate, test, and recommend equipment that should be carried on rescue/USAR apparatus. Generally speaking, these standardized lists are intended as templates, or starting points, for fire/rescue departments to outfit their apparatus with the right amount and type of equipment to get the job done based on statewide and regional hazards and needs.

For example, in California this concept was implemented beginning in the 1970s with standard typing of wildland firefighting resources as part of the evolution of the ICS. Engine companies, helicopters, bulldozers, and firefighting crews were typed according to their capabilities and characteristics. Each type (e.g. Type I engine company capable of road-based wildland operations and structure protection, Type I helicopters based on size, personnel capacity, and water-carrying capacity, etc) is required to carry a certain complement of equipment to do the job that will be assigned to it.

All the typing information (including minimum equipment and staffing) is consolidated in various manuals, as well as the ICS Field Operations Guide (FOG), and carried on every fire department vehicle and apparatus in the state of California. This allows the IC and other officers to quickly match the needs of their incident with the standard typing of resources that they can call upon. This is especially important during major disasters when unusual types of resources may be required.

In California, with its long history of rescue-related emergencies and disasters, this concept has been extended to include standardized typing of USAR resources, including:

- USAR companies (apparatus-equipped and staffed with crews trained to a certain level)

- USAR crews (personnel without the apparatus who can be called in to augment existing crews and apparatus at the scene of a disaster)

- USAR task forces (based on the FEMA USAR task force standards)[2]

- swift-water rescue task forces[3]

To offer an example of how typing works in states like California, appendix I is a listing of standard equipment inventories for USAR resources.

Equipment inventory for USAR and rescue companies[4]

The equipment carried on or immediately available to rescue companies and USAR companies varies according to the mission of the unit and the hazards/conditions found in the unit's jurisdictional area (Fig. 4–18). Equipment requirements are also dependent on local policy and protocol. The judgment and preferences of personnel assigned to the units (often guided or influenced by agency-specific equipment development committees) may also determine equipment needs. Other determining factors are the wishes of supervisors; history and habit;[5] local, regional, state, and federal standards; and of course the budget process.

Fig. 4–18 All-terrain Vehicles Like This Are In Use by Several FEMA USAR Task Forces

Finally, equipment inventories for USAR and rescue companies are strongly dependent on the configuration size of the rescue/USAR apparatus. Some heavy rescue apparatus are designed to accommodate more than others. In addition, we all know that as the years go by and new hazards are recognized, firefighters and rescuers have a tendency to fill every last nook and cranny with whatever equipment will fit in the apparatus. Obviously it's impractical to list every possible item here or to apply the standards used in one region to the requirements of another region that may have different or unique hazards (Fig. 4–19).

Fig. 4–19 The New Jersey State USAR Task Force Uses This Fifth-wheel Apparatus

However, it is sometimes helpful to review the equipment lists of USAR and rescue companies of various fire/rescue agencies to compare notes and determine whether equipment specified in some regions may also be applicable to other regions.

With that in mind, appendix I is a sample of equipment carried by USAR companies of the LACoFD. These companies in combination with the USAR engine assigned to the same fire stations are considered USAR task forces in the LACoFD USAR response system.[6] Appendix II is the standard equipment inventory for California-based USAR resources.

The reader will note first that the equipment inventory of these units exceeds that of a Type I California USAR company. This should not be considered unusual or extraordinary because the state standard is designed as a template for the *minimum* complement of equipment to do the job required of the typical USAR company in California. It has always been recognized and expected that some fire/rescue agencies will carry inventories that exceed the standard, based on the hazards and needs of their jurisdictional/response area (Fig. 4–20).

Fig. 4–20 An Existing Vehicle Adapted to Carry Specialized Equipment for Short Distances

In this case, the LACoFD equipment inventory is intended to support the USAR company/task force firefighters in their designated mission to conduct a wide range of technical SAR operations. This includes operations that occur in the city, mountains, ocean, and deserts of L.A. County. In the case of mutual aid responses, it includes all of southern California.[7] It should be noted that this is not an endorsement of any particular brand or manufacturer of tools and equipment. The market is wide open with both established and emerging manufacturers and suppliers, and the best advice is for fire/rescue agencies to explore the full range of available options.

Appendix I: Sample USAR Company Equipment Inventory

This list is an inventory of the tools and equipment used by the LACoFD (Fig. 4–21).

Fig. 4–21 Semi with Flat-bed Trailer for Transporting Heavy Equipment

Equipment inventory list

- Camera, thermal imaging system
- Rotary hammer bit kit
- Radio, hand-held, intrinsically safe, command frequency
- Wrench, torque
- Technical search camera, snake-eye
- Radio, hand-held, intrinsically safe, tactical frequencies
- Axe
- Tote set, 2-piece
- Air bag, rescue, 21.8-ton, 11.1-in. lift

- Air bag, rescue, 17-ton, 9.2-in. lift
- Air bag, rescue, 12-ton, 8-in. lift
- Chains, tire
- Drill, wood $1/2$-in. x 18-in.
- Wrench, impact $3/4$-in. drive
- Reducer, $2^1/2$-in. to $5/8$-in.
- Chain, cutting, diamond
- Drill, wood $3/4$-in. x 18-in.
- Air Bag, rescue, 43.8 ton, 15.5-in. lift
- Driver, screw, flat, 10-in.
- Wrench, spark plug
- Chisel, pneumatic
- Air bag, rescue, 1.5-ton, 3-in. lift
- Air bag, rescue, 73.4-ton, 20-in. lift
- Air bag, rescue, 35-ton, 10-in. lift (rectangle)
- Air bag, rescue, 3.2-ton, 3.5-in. lift (rectangle)
- Air bag, rescue, 4.8-ton, 5-in. lift
- Drill, pneumatic, $3/8$-in.
- Rotary hammer and charger
- Chains, tire, snow
- Chain, tensioner
- Arcair® rods, box, $3/8$-in.
- Chairs, folding, for confined-space rescue equipment donning, standby team readiness, and personnel rehab
- Coupling, set
- Saw, rotary, blade, 16-in., concrete, dry
- Ram set bolts, 7-in. x $1/2$-in.
- Ram set nails, 3.5-in.

- Saw, rotary, blade, 14-in., concrete, wet
- Ram set bolts, 9-in. x 1-in.
- Ram set bolts, 7-in. x $3/4$-in.
- Ram set bolts, 5.5-in. x $1/2$-in.
- Saw, rotary, blade, 16-in., diamond
- Saw, rotary, blade, 14-in., concrete, dry
- Saw, rotary, blade, 16-in., concrete, wet
- Saw, rotary, blade, 16-in., carbide tip (wood)
- Saw, rotary, blade, 16-in., metal-cutting
- Blade, woodcutting, 16-in.
- Boat motor, 25-hp outboard with fuel tank
- Supplied air breathing apparatus (SABA) air hose, 100-ft/sec
- Saw, rotary, 14-in., K-700
- Fiber optic technical search system
- IRB
- Drill/hammer drill, 24-V
- Pneumatic rescue system (Airgun 40)
- Bit, pilot point set, 19-piece
- EMS backpack for mountain rescue, collapse, confined-space, and other limited access patient treatment
- Chuck, $1/2$-in., with key for DW5351
- Map books and maps for the following counties: L.A., Orange, Ventura, Santa Barbara, Riverside, San Bernardino, San Diego, San Luis Obispo, Kern (jurisdictional and regional counties)
- Map books and maps for the following states: California, Arizona, Nevada, and Oregon (jurisdictional and adjacent states)
- Adaptor, chuck SDS

- Building triage/marking kit

- Rotary hammer, $1^1/_2$-in., kit, electric

- Stanley rescue system parts kit

- Bolt cutter, 36-in.

- Pump, hand-fuel

- Bolt cutter, 12-in.

- Life detector technical search system and accessories

- Hose, garden

- Power unit hydraulic, Stanley-HP1 rescue system

- Chainsaw, diamond, Stanley rescue systems, hydraulic, DS11

- Saw, skill, worm drive $7^1/_4$-in.

- Cut-off saw, Stanley rescue system, 14-in., hydraulic, CO23

- Technical search device, Olympus fiber-optic

- SABA air cart air supply system

- Breaker, 90-lb, Stanley rescue system, BR89

- Bit, moil-point, Stanley rescue system breaker

- Pump, trench shore

- Ducting, ventilation, accordion, 15-ft

- Chainsaw, gas

- Half-backboard, LSP (Miller half-back), for confined-space vertical extraction and other limited-access vertical rescues

- Saw, reciprocating, 18-V Rechargeable

- Adaptor, rotary hammer, spline to tapered shank

- Saw, rotary, 16-in., Partner K-1200

- Kit, pneumatic shoring feet, in hardened carrying case

- Chain kit, for combo hydraulic spreader

- Heavy-duty lockout tags, package

- Air horns, hand-held

- Cutting torch, oxy-acetylene

- Tool, release, trench shore

- 4:1 Mechanical advantage rope system, pre-rigged, in high-angle rescue packs

- Fluid, hydraulic, trench shore

- Rotary hammer bit, core bit adapter

- Hydraulic rescue tool system power unit

- Ice chest for biopack SCBA canisters and rehab

- Kit, air bag, rescue, low-pressure

- Ram, hydraulic rescue system, 20-in.

- Hose, airbag, rescue, 16-ft, black

- Ram, hydraulic rescue system, 30-in.

- Wrench, pipe, 14-in.

- Ram, hydraulic rescue system, 60-in.

- Air bag, rescue, master control package

- Breaker, hammer, 60-lb

- Resuscitator with airways, bag valve mask, etc., in backpack for mountain rescue, vehicles over-the-side, confined-space rescue, and other limited-access patient treatment

- Saw parts and repair kit, in ammo box

- Atmospheric monitor kit, hardened case, with organizer

- Confined-space communications kit, in hardened case

- Saw, reciprocating, 24-V

- Lock-out tag-out kit, in hardened case

- Drill, hammer, rotary

- Extinguisher, dry chemical

- Bit, wood-eater, 5-piece set

- Ear plugs, package

- Chainsaw, electric, 16-in.

- Spinner, rivet

- Rescue frame

- Burn sheet, set

- Kit, trailer hitch for towing PWC trailer

- Dirt vacuum system

- Air knife system

- Clipboard, technical rescue, with technical rescue ICS worksheets and other incident documentation items

- Swift-water rescue hose inflator system

- Swift-water rescue line-thrower system

- Chainsaw bar, 20-in.

- Chainsaw chain blade, 16-in., carbide tip

- Swift-water rescue board

- Meals ready to eat (MRE), case, for disaster operations and overnight hike-in rescues in the mountains

- Confined-space scene-management kit, case

- SCBAs, with masks (one per post position)

- Night-vision goggle systems

- Hammer, sledge, 16-lb

- Radio, hand-held, mountain rescue frequencies

- Confined-space communications talk box

- Rope, rescue, $7/_{16}$-in., 200-ft

- Rope, rescue, $1/_2$-in., 250-ft

- Water, bottled, case, for disaster operations, long-term rescues, and rehab

- Kit, IRB

- Confined-space communications, operator external mic/mute switch, 1-ft

- RIC pack, air with bottle, to supply emergency air to trapped firefighters (with or without their SCBAs on) and to trapped citizens during fireground or rescue scene extraction

- Hook, grappling

- Rapid intervention straps, CMC, one per post position

- Tripod rescue system, for cliff rescues, deep-shaft rescue, confined-space vertical entry, rooftop operations on high-rise rescues, etc.

- Alarms, personal safety system, one per post position

- Rescue straps, Cearley straps for helo/swift-water, marine rescue, and cliff rescue operations

- Helicopter high-rise rescue pack

- High-rise officer's hose pack

- Litter, rescue, Junkins

- Litter, rescue, Ferno, for PWC rescues

- Breaker, Bosch, hand-held

- Confined-space communications command module

- Atmospheric monitor calibration kit

- Bolt cutter, 42-in.

- Dikes, crosscut

- Pliers, needle-nose

- Pliers, lineman

- Litter with pre-rig

- Chain, 20-ft

- Ram bar, sliding

- Technical rescue system search camera

- Shovel, round-point, 3-ft

- Shovel, long-handle, square-point

- Night-vision monocle system

- Atmospheric monitor, sampling pump

- Ram set, powder-actuated nail gun

- Kit, tripod, pneumatic

- Wrench, crescent, 8-in.

- Pneumatic shoring system

- Lifeguard rescue can for surf, marine, lake, and swift-water rescues

- Wrench, crescent, 12-in.

- Blanket, rescue, reflective

- Confined-space communications, operator external mic/mute switch and cord, 20-ft

- Set, index, drill

- Litter, confined-space rescue (SKED sled)

- Victim rescue harnesses, 4 needed

- Probe, rescue, telescoping, 6-ft, stainless steel

- Spray paint, fluorescent orange, case, for structural- and victim-marking systems

- Combination tool, hydraulic spreader/cutter

- Anchor bolt, box, $1/2$-in. x 7-in.

- Ram, rescue, 40-in.

- Anchor bolt, box, $3/4$-in. x 7-in.

- Anchor bolt, box, 1-in. x 9-in.

- Rebar cutter, hand-pump

- Rope, rescue, 600-ft, 2 sets

- Saw, beam, 16 $1/4$-in.

- Bit, concrete, $3/8$-in., L-10 depth 10

- Bit, concrete, $1/2$-in., L-10 depth 10

- Clipboard, medical, for EMS documentation

- Bolt cutter, 14-in.

- Pneumatic shore-regulator

- Electrical adaptors, various, in kit

- Set, wrench, ignition

- Electrical adaptors

- Bit, concrete, 1-in., L-10 depth 10

- Rope bags, for all ropes carried in USAR company

- Sprocket, drive, chainsaw, Stanley rescue system

- Bio-pak closed-circuit SCBAs, one per post position, for mine/tunnel rescue, high-rise fires, helo/high-rise operations, long-distance confined-space operations, etc.

- Dive rescue kit, one per post position

- Ice and snow rescue kit, one per post position

- Avalanche poles, kit

- Pump, water, Stanley rescue system chainsaw

- Wrench, crescent, 10-in.

- Wrench, combination, $1 1/4$-in., Stanley

- Wrench, combination, $1 1/16$-in., Stanley

- Plug, spark, Stanley rescue system power unit

- Pre-rig O-rings, steel

- Saw, concrete, circular, Stanley rescue system

- Electrical adaptor

- Filter, air, power unit, Stanley rescue system

- Rope, utility, $1/2$-in., 200-ft, in bag

- Filter, oil, Stanley rescue system

- Pulleys, knot pass, 2-in., in high-angle rescue packs

- Bar, chainsaw, Stanley rescue system

- Ducting, ventilation, accordion, 25-ft, 2 sets, for confined-space and mine/tunnel rescues

- Rebar cutter, electric, 1-in. capacity

- Pre-rig systems for litters
- Rotary hammer, battery-powered
- Oil, hydraulic, 5 gals, Stanley rescue system
- Blade, rotary, metal, 16-in.
- Respirator, dust and mist
- Prusiks, long, in high-angle rescue packs
- Edge pad, canvas, in high-angle rescue packs
- Rope, anchor, 25-ft, in high-angle rescue packs
- High-angle rescue pack, industrial, steel hardware
- Edge rollers, in high-angle rescue packs
- High-angle rescue pack, pony pack, mountain, aluminum hardware
- Rescue platforms, 10-person marine rescue platforms, for helicopter deployment into ocean during marine disaster operations
- Sawzalls, with hardened cases, and blades, electric
- Search and recon kit, in backpacks
- Plumb bob
- Rods, cutting, 50/box, 4 boxes needed, for Arcair® rescue metal cutting system
- Brooms, corn
- Pressurized water spray can, Hudson-type
- Anchor bolt, box, $1/_2$-in. x $5 1/_2$-in.
- Air bag, rescue, hose, 16-ft, yellow
- Air bag, rescue, dual controller
- Chainsaw chain, 20-in., carbide tip, spare
- Funnel, $7 1/_2$-in. diameter
- Chainsaw chain, 16-in., standard, spare
- Cutting torch cylinder, oxygen

- Axes, flathead
- Cutting torch cylinder, acetylene
- Kit, chain breaker, Stanley
- Axes, pick-head, with belt, one per post position
- Helmet, victim, rescue
- Wrecking bar, 3-ft
- Filter, hydraulic, Stanley rescue system
- Air bag, rescue, hose, 16-ft, black
- Bit, chisel-point, Stanley rescue system, breaker
- Pliers, vice-grips
- Digital camera, still
- Wrench, Allen, 15-piece set
- Come-Alongs, $1/_2$-ton
- Kootenay carriages, in high-angle rescue packs
- Straps, 2-in. x 20-ft, tie-down
- Hoses, hydraulic, Stanley rescue system, twin, 50-ft
- Air horn canister, spare
- Bars, digging, 16-lb, 72-in.
- Hydraulic jacks, 12-ton, bottle
- Devices, friction, 8 plate, aluminum, in high-angle rescue packs
- Air bag, rescue, hose, 16-ft, red
- Pneumatic adaptors, kit
- Pneumatic shore, 36.3-in. to 58.0-in.
- Hydraulic adaptors, kit
- Rope, rescue, 200-ft
- Rope, rescue, 300-ft
- Shovel, long-handle, round-point

- Electrical adaptor kit

- Prusiks, short, in high-angle rescue packs

- Chisel, cold, $^7/_8$-in. x 11-in., SDS Max

- Collection plate, aluminum, in high-angle rescue packs

- Bank chargers, battery, various

- Radio, hand-held, USAR, talk-around entry frequency

- Tygon tubing, 1 each: 25-ft, 50-ft

- Point, bull, 12-in., SDS Max

- Cribbing, 4 x 4 x 24-in. kits

- Salvage covers

- Oil, engine, 2-cycle, 8-oz

- Radio, hand-held, aviation

- Pick-off straps, in high-angle rescue packs

- Sigg bottle, 1-qt

- Swift-water rescue equipment packs

- Sprocket, nose, bar, Stanley rescue system

- Bags, drop, one per post position

- Brady single-pole circuit breaker lockout, package

- Brady multi-pole lock-out

- Shores, speed, trench shores

- Arcair® rods, box, $^1/_4$-in.

- Lights, portable, ground, 500-W

- Charger, 1-hr, 3-stage charging for 24-V DeWalt cordless

- Atmospheric monitors, one per post position

- Sawzall blades, package, metal

- Sawzall blades, package, wood

- Pneumatic shore, 18.8-in. to 24.5-in.

- Screwdriver, common

- Radio batteries, VHF and UHF, charged

- Radiation monitoring kit, civil defense

- Vests, orange, position assignment, kit

- Pneumatic shore, 55.5-in. to 87.3-in.

- Underwater radio kit

- Swift-water strobe lights, on each personal flotation devices (PFD)

- Hydraulic rescue system hose, 20-ft/sec

- Pneumatic shore, extension, 36-in.

- Tape, duct, box

- Device, friction, 8 plate, steel, in high-angle rescue packs

- Carabiner, extra-large, steel, in high-angle rescue packs

- Confined-space communications cable, 200-ft, with connectors

- Helmets, helicopter operations, one per post position

- Headlamps, Pelican Versa Bright II, one per post position

- Confined-space rescue communications cable, 50-ft, with connector

- Edge protection, Ultra Pro, in high-angle rescue packs

- Confined-space rescue communications, face mask rescue set, with speaker

- Bit, $1^1/_2$-in., wood-eater

- Confined-space rescue communication cable, 100-ft, with connector

- Personal watercraft, 2 needed, for river and flood rescue operations and for submerged-victim search operations

- Bit, $1/2$-in., straight shank

- Trailer, PWC, rescue

- Bit, $3/8$-in., straight shank

- SABA escape bottles, spare, 10-minute, one per post position

- Swift-water PFD, one per post position, one for victim

- Light, hand, King Pelican

- Rotary hammer, Hilti, battery-powered, 36-V

- Fuel can, 1-gal, safety

- Swift-water knives, one per PFD

- Horse rescue harness, large

- SABA mask kit, case

- Mask, swim, one per post position

- Harness, full-body, Class 3, one per post position

- Pulleys, 4-in., in high-angle rescue packs

- Blade, rotary, concrete, 16-in.

- Pneumatic shore, extension, 12-in.

- Pneumatic shore, 24.8-in. to 35.4-in.

- Swift-water throw bags, with 70-ft rope

- Radiation pager, one per post position

- Cordage, utility 10M, in high-angle rescue packs

- Blade, diamond continuous rim, 16-in.

- Horse rescue harness, medium

- Dog rescue harness

- Fins, swim, pair, one per post position

- Whistles, on each PFD

- Cyalume light sticks, boxes with different colors

- GPS, hand-held

- Saw, electric, 16-in., beam

- Saw, circular, electric, $10 1/4$-in.

- Saw, rotary, blade, 14-in., diamond

- Air chisel/impact wrench, pneumatic, kit

- Air bag hose, green, 32-ft

- Distance measuring device

- Grip hoist, TU28, with 50-ft cable

- Handtruck

- Air bag hose, blue, 32-ft

- Mechanical grabber, electrical line

- SABA air-hose reel, with 250-ft air line, victim

- Lumber crayon, red, box

- Shovel, scoop, D-handle

- Parts-cleaning station

- Parts-cleaning station flow-thru brush

- Air bag hose, 32-ft, red

- Confined-space rescue communications boom mic wind screen

- Lumber crayon, yellow, box

- Nut driver set, standard

- Search camera reel system, 300-ft with standard camera/two-way communication head, for observation and communication in deep-shaft, mine/tunnel, confined-space, and structural collapse rescues, etc.

- Screw driver bit set, 60-piece

- Ratchet, standard tooth, flexible head, $3/8$-in.

- Ratchet, $5^1/_2$-in., stubby, flexible head, $3/_8$-in.

- Punch and Chisel Set, 24 pc.

- Pliers, lineman

- Pliers, arc-joint

- Pliers set, reach needle-nose

- Pliers set, locking

- Short-arm hex set, metric, 11-piece

- Short-arm hex set, standard, 11-piece

- Nut driver set, metric

- Multi-meter, digital LCD

- Handle, spinner, $1/_4$-in.

- Handle, flex T, $1/_2$-in.

- Wrench set, adjustable, 5-piece

- Respirator canister, replacement, case

- Chainsaw bar, 16-in.

- Electric detection device, hot stick

- Respirators, APR, one per post position

- Pliers set, pro

- Wrench, combo, metric, 23-mm

- Anchor plate, hanger, stainless steel, box

- Binoculars, 10 x 25

- Ram, hydraulic rescue, accessory kit

- Slings, 2-in. x 20-ft, in high-angle rescue packs

- Label machine

- Brady Plug Lockout

- Wrench, combo, standard, $1^5/_{16}$-in.

- Wrench, combo, metric, 6-mm

- Screw driver set

- Wrench, combo, metric, 25-mm

- Backboard, with straps (Miller board)

- Wrench, combo, metric, 20-mm

- Wrench set, combo, metric, 12-pt, 17-piece

- Wrench set, combo, 12-pt, 17-piece

- Socket wrench set, 235-piece

- Socket tray set, 3-piece

- Socket set, bit, 19-piece

- Wrench, combo, metric, 26-mm

- Anchor eye nut, $1/_4$-in.

- Saw, rotary, kit

- Level, 4-ft

- Drill, variable speed, $1/_2$-in., Dewalt

- Drill bit set, steel, $1/_8$-in. to $5/_8$-in.

- Drill bit set, carbide, $1/_4$-in. to $5/_8$-in.

- Die grinder, pneumatic, kit

- Water, drinking, case

- Laser pointer

- Dosimeters, one per post position

- Anchor Eye Nut, $3/_8$-in.

- Nail kit

- Anchor eye nut, $1/_2$-in.

- Awning anchor kit, heavy-duty

- Video camera, with light, spare battery, in case

- Awning heavy-duty leg set, EZ-UP, for rehab and equipment pool

- Awning, EZ-UP, for rehab and equipment pool

- Level A entry suits, one per post position, for WMD terrorist attacks, potentially contaminated structural collapse sites, etc.

- Metal detector
- Stanley breaker, 45#, hyd., BR45
- Ladder, Little Giant
- Anchor eye nut, $5/8$-in.
- Search cam video transmitter and connector
- Fuse, hwy, box
- Hall runners
- WMD terrorism antidote kit
- Nail gun nails, case 8-pack
- Nail gun nails, case, 16-pack
- Nail gun, pneumatic, kit
- Tow straps, nylon
- Tool chest
- Hose, $1^{3}/4$-in., 50-ft/sec
- Mechanical axe, electrical-line cutter
- Ducting, Lay-Flat L
- Respirator parts kit
- Swift-water rescue briefcase with logbook, waterway rescue preplans, inundation maps for dam failure and fast-rise flooding; tsunami inundation maps for coastal zones vulnerable to tsunamis
- Pickett anchor set, in backpack for mountain rescues, mine/tunnel rescues, swift-water rescues, etc.
- Breaker bit, bull-point, $1^{1}/8$-in. hex
- Breaker bit, chisel, 1-in., $1^{1}/8$-in. hex
- Chainsaw kit
- Air bag, rescue, hose, black, 32-ft
- Air bag, rescue, hose, 32-ft, yellow
- Rope, tagline, 300-in., 8-mm in bag
- Screwdriver, straight
- Exothermic cutting torch

- Pliers, pair, slip-joint, 10-in.
- Target hazard fire attack preplans
- SABA air hose kit, case
- Funnel, $3^{1}/2$-in. diameter
- Hydraulic rescue tool pump, manual, with foot pedal
- Ground-penetrating radar, technical search, field-ready system
- Rotary hammer bit, 1-in.
- Ventilation blower, electric, intrinsically safe
- Ventilation blower conductive duct, 8-in. x 4-ft
- Tape, electrical, black, roll
- Breaker, chain
- Ventilation duct coupler
- Rotary hammer bit, coring
- Rotary hammer bit, $3/8$-in.
- Pipe cutter, multi-wheel
- Hacksaw, heavy-duty, with 10 blades
- Crosscut hand saw
- Camera, digital
- Ventilation blower, 12-in. duct adaptor
- Electrical junction box, 4 outlet, with GFI
- Level, line
- Electrical adaptor, Wye
- Thermal imaging system, hand-held, with video transmitter link
- Saw, skill, blade, $7^{1}/4$-in. carbide tip
- P-S wave detectors, for early warning of incoming earthquakes and aftershocks during collapse rescue operations
- Haligan

- Smart Levels

- Ventilation duct storage bag

- Sledge hammer, 10-lb

- Bauman bag vertical extrication C-spine immobilizer

- Oxygen bottle, 55-cu ft

- Ram set studs, box, $^3/_8$-in.

- Tape measure, 25-ft

- Nail puller, Catpawels

- Brake bars, in high-angle rescue packs

- Hammer, framing, 24-oz straight-claw

- Sledge hammer, 3-lb

- Ram set booster, red, Hilti

- Dewatering pumps

- Ram set booster, yellow, Hilti

- Cutting torch cutting tip #2

- Camera, digital, motion picture

- Carpenters pencil, box

- Cutting torch cutting tip #000

- Exothermic torch $^3/_8$-in. conversion kit

- Compasses, Type 3, one per post position

- Cutting torch cutting tip #1

- Cutting torch welding tip #00

- Cutting torch tip, multiflame #8

- Cutting torch tip, multiflame #6

- Cutting torch cutting tip #00

- Saw, rotary, blade, 14-in., carbide tip (wood)

- Exothermic torch washers

- Carabiners, large, locking, aluminum, in high-angle rescue packs

- Electrical cords, 50-ft

- Carabiners, large, locking, steel, in high-angle rescue packs

- Ram set fasteners, 2.25-in. x $^3/_8$-in.

- Ram set fasteners, 1.75-in. x $^7/_{16}$-in.

- Wedges, pair, 6 x 6 x 30-in. case

- Wedges, pair, 6 x 2 x 18, case

- Lumber, plywood, 1-in. and 2-in.

- Carabiner, standard, locking, aluminum

- Wedges, pair, 2 x 4 x 18

- Palm pilot for IT applications

- Picket anchors, 1-in. x 4-ft, cold rolled steel

- Webbing, green, 5-ft, in high-angle rescue packs

- Screw jacks, $1^1/_2$-in. x 18-in., with foot

- Cribbing, 6 x 6 x 30, cases

- Stick, light, 12-hr

- Air bag, rescue, pneumatic components kit

- Air bag, rescue, 31.8-ton, 13.1-in. lift

- Auxiliary electrical adaptors, kit, assorted

- Electrical adaptor, Wye

- Chainsaw air filter

- Saw, rotary, air filter

- Cutting torch wrench, 8

- Label machine tape, 4-in.

- Saw, rotary, drive belt, spare

- Rotary hammer bit, $1^1/_4$-in.

- Rotary hammer bit, $^7/_8$-in., spline shank

- Awning Side Wall, EZ-UP—For Rehab and Equipment Pool

- Level, 6-in.

- Rebar cutter blades

- Engraver, electric

- Lineman gloves, pair

- Pliers, tongue-and-groove, 14-in.

- Cell phones, one per post position, for emergency communications

- Screwdriver, Phillips

- Nails, box, 25-lb, 8-p

- Nails, box, 25-lb, 16-p

- Adapter kit, air knife and air vacuum, to provide compressed air for SCBA bottles for victims trapped in quicksand, mud flats, and trench/excavation collapses remote from USAR apparatus

- Screwdriver, Phillips, stubby

- Chalk-line chalk, 8-oz

- Rebar cutter, battery

- Rebar cutter, battery, spare blades

- Screwdriver, slotted, stubby

- Horse rescue-harness, small

- Animal rescue-snare

- Rotary hammer bit, $^{11}/_{16}$-in., spline shank

- Electrical cord reel, 10

- Nails, box, 25-lb, 16-p duplex

- Etriers, in high-angle rescue packs

- Crane operators manual

- Ram set boosters, purple, Hilti

- ICS FOG

- Battery, 12-V lead acid, spare for exothermic cutting torch and other similarly powered tools

- Ventilation blower ducts, 8-in., 1 each: 25-ft, 15-ft

- Bars, pry- and pinch-point, 60-in., 18-lb

- FEMA USAR National Response System FOG

- Rescue resource list with contact numbers for local/state/federal response agencies, OSHA, public works, crane operators, structural engineers, demolition contractors, etc.

- Cribbing, 2 x 4 x 24-in. cases

- Wedges, pair, 4 x 4 x 24-in. cases

- Generator, 5-kW, portable, gas

- Swift-water helmets

- Fireground search rope pack

- Blade, carbide tip, 14-in. fire rescue

- Ellis jacks

- Lumber, 6-ft x 6-ft x 20-ft

- Lumber, 2-ft x 12-ft x 12-ft

- Petrogen metal-burning systems

- Cans, gasoline, 1-gal, for Petrogen metal-burning systems

- Cribbing, 2 x 6 x 30-in. case

- Ram set pins, box, steel, $^{7}/_{8}$-in.

- Osborn aluminum lockouts, 1-in. hasp

- Osborn aluminum lockouts, $1^{1}/_{2}$-in. hasp

- Ram set pins, box, $1^{7}/_{8}$-in.

- Ram set pins, box, $2^{7}/_{8}$-in.

- Anchor kit, in ammo boxes, one per size

- Communicable disease kits, one per post position

- Chock block, metal

- Line, chalk

- Eight plate, rescue, aluminum, in high-angle rescue packs

- Eight plate, rescue, steel, in high-angle rescue packs

- Carpenter belts, one per post position

- Hydraulic tool fluid, 3 gal needed

- Square, tri/speed

- Shovels, folding

- Rope, rescue, 100-ft

- Chalk line

- Lumber, 2-ft x 4-ft x 20-ft

- Lumber, 4-ft x 4-ft x 16-ft

- Respirator dust pre-filter

- Haul buckets, metal, 5-gal

- Light saddle, Pelican

- Electrical adaptor

- Diagonal cutters

- Swift-water dry suit, one per post position

- Electrical plug, L14P, male plug

- Swift-water goggles, one per post position

- Swift-water gloves, pair, one per post position

- Spark plug, for rotary saw

- Swift-water fins, pair, one per post position

- Kneepads, pair, Cordura, with plastic cover, one per post position

- Elbow pads, pair, Cordura, with plastic cover, one per post position

- Swift-water booties, wet suit, pair, one per post position

- Atmospheric monitor, battery shell pack

- SABA air-hose reel, with 250 ft of air line and communications line

- Electrical plug, L14R, female receptacle

- Spark plug, for Stihl 044 chainsaw

- Economy lockout padlock with $2^1/_2$-in. shackle

- Webbing, blue, 12-ft

- Floodlight, 500-W, portable

- Atmospheric monitor, spare batteries, lithium

- Canvas bag, tool, carpenters, extra-strength, one per post position

- Saw, rotary, blade, 14-in., metal-cutting

- Ellis clamps

- Filter, water/dust stop

- Paddles, Carlson, for IRB operations

- Radio harness, chest, one per post position

- Exothermic cutting torch collet nuts, $^1/_4$-in.

- Exothermic torch flash arrester

- Zip ties, medium-duty, 10-in.

- Radio bone microphones, for Motorola SABA radio, one per post position

- Topographical maps for mountainous areas in jurisdictional area/surrounding regions

- Pipe, schedule 40, $1^1/_2$-in. x 20-ft

- Webbing, red, 22-ft, in high-angle rescue packs

- SCBA bottles, 60-min, one primary and one spare per post position; five extra for equipment operations

- Pulleys, Prusik-minding, in high-angle rescue packs

- Electrical plug, L5P, male plug

- Post screw jacks

- Electrical plug, L5R, female receptacle

- Pipe, steel schedule 40, $1^1/_2$-in. x 4-ft

- Air bag, rescue, regulators, in hardened case for industrial and street operations

- Pulleys, 2-in., in high-angle rescue packs

- Air bag, rescue, regulators, in backpack for rapid intervention operations and collapse rescue operations

Appendix II: Sample USAR Operational Capability Minimum-Equipment Lists

These lists identify the minimum number of tools and amount of equipment needed to provide a safe and acceptable level of service for each of the four levels of USAR operational capability. The amount, size, and type of equipment listed can be increased to provide a higher degree of safety and service in each level of USAR operational capability (Fig. 4–22).

Fig. 4–22 An Example of Equipment being Carried on Modern USAR and Rescue Units

USAR basic level

The following is a list of the minimum equipment required for a USAR basic level operation:

- Two 8- to 10-lb sledge hammers
- Two 3- to 4-lb sledge hammers
- Two cold chisels (1-in. x $7^7/_8$-in.)
- Four pinch point pry bars (60-in.)
- Two claw wrecking bars (3-ft)

- Two hacksaws (heavy-duty)
- Three carbide hacksaw blade packages
- Two crosscut handsaws (26-in.)
- One cribbing and wedge kit (see appendix III for tool/kit descriptions)
- One first-aid kit (See appendix III for tool/kit descriptions)
- One trauma kit (See appendix III for tool/kit descriptions)
- Two blankets (disposable)
- One backboard with 2 straps
- One bolt cutter (30-in.)
- One scoop shovel, D-handle
- One building marking kit (see appendix III for tool/kit descriptions)
- One axe (flat-head)
- One axe (pick-head)

USAR light level

The following is a list of the minimum equipment required for a USAR light level operation:

- One USAR basic equipment inventory
- Two 150-ft x $^1/_2$-in. Kern mantles, static, NFPA-approved
- Two friction devices (see appendix III for tool/kit descriptions)
- Twelve carabiners (locking D, 11-mm)
- Six camming devices (see appendix III for tool/kit descriptions)
- Three pulleys, rescue (2-in. or 4-in.)
- One litter and complete pre-rig (see appendix III for tool/kit descriptions)
- One webbing kit (see appendix III for tool/kit descriptions)
- Two edge protections (see appendix III for tool/kit descriptions)

- Two pick-off straps (see appendix III for tool/kit descriptions)

- Two commercial harnesses (Class II or better)

- Six steel pickets (1-in. x 4-ft)

- Two 3- to 4-lb short sledge hammers

- One chainsaw (see appendix III for tool/kit descriptions)

- Three tape measures (25-ft)

- One shovel (long-handle, square-point)

- One shovel (long-handle, round-point)

- Two framing hammers (24-oz)

- Two tri or speed squares

- Two carpenter belts

- One set of nails (see appendix III for tool/kit descriptions)

- Two hydraulic jacks (minimum 5-ton)

- Two rolls duct tape

USAR medium level

The following is a list of the minimum equipment required for a USAR medium level (Fig. 4–23) operation:

Fig. 4–23 Tall Apparatus Sometimes Requires Step Platforms, Slanted Shelves, and Other Customized Features to Afford Quick Access

- One USAR basic equipment inventory and light equipment inventory

- One air bag set (3 bag, 50-ton with 3 spare air bottles)

- One bolt cutters (heavy-duty, 42-in.)

- One generator (5-kW)

- Four floodlights (500-W)

- Six extension cords (50-ft)

- One junction box (4-outlet with GFI)

- One Wye electrical adapter

- One circular saw (12-in., with $2^1/_2$ gal fuel)

- Two circular saw blades (12-in., carbide tip)

- Twelve circular saw blades (12-in., metal-cutting)

- Two circular saw blades (12-in., diamond, continuous rim)

- One pressurized water spray can

- One rotary hammer ($1^1/_2$-in.)

- One rotary hammer bit kit (see appendix III for tool/kit descriptions)

- One anchor kit (see appendix III for tool/kit descriptions)

- One saw, electric ($10^1/_4$-in.)

- Two skill saw blades ($10^1/_4$-in., carbide tip)

- Twelve skill saw blades ($10^1/_4$-in., metal-cutting)

- One sawzall

- Twelve sawzall blades (wood)

- Eighteen sawzall blades (metal)

- Two ropes (300-ft x $^1/_2$-in., static kern mantle NFPA-approved)

- Two ropes (20-ft x $^1/_2$-in., static kern mantle NFPA-approved)

- Three pulleys, rescue (2-in. or 4-in.)

- Two friction devices (see appendix III for tool/kit descriptions)

- Twelve carabiners (locking D, 11-mm)

- One webbing kit (see appendix III for tool/kit descriptions)

- One etrier set

- Two commercial harnesses (Class II or better)

- Two shovels (folding, short)

- Four haul buckets (metal or canvas)

- Eight Ellis clamps

- One Ellis jack

- Eight 4-ft x 4-ft x 8-ft lumber

- Six screw jacks (pairs, $1\frac{1}{2}$-in.)

- One pipe cutter (multi-wheel, $1\frac{1}{2}$-in.)

- Six pipes (6-ft x $1\frac{1}{2}$-in., Schedule 40)

- Two hi-lift jacks with extension tubes

- One cribbing and wedge kit (see appendix III for tool/kit descriptions)

- One Come-Along (2/4-ton)

- One chain set (see appendix III for tool/kit descriptions)

- One tool kit (see appendix III for tool/kit descriptions)

- One demolition hammer, small (see appendix III for tool/kit descriptions)

- One demolition hammer, large (see appendix III for tool/kit descriptions)

- One electrical detection device (see appendix III for tool/kit descriptions)

- One ventilation fan (see appendix III for tool/kit descriptions)

- One 3-range air monitor

USAR heavy level

The following is a list of the minimum equipment required for a USAR heavy level (Fig. 4–24) operation:

Fig. 4–24 A Wide Array of Equipment and PPE for Rescue Companies and USAR Units

- One USAR basic equipment inventory

- One USAR light equipment inventory

- One USAR medium equipment inventory

- Six SCBA (with personal alarm and 1 spare bottle each)

- Three SABA

- Umbilical system with escape bottles and 250-ft hose each

- One 3-range air monitor

- One Tri Pod (human-rated, 7-ft to 9-ft with hauling system)

- Two full-body harnesses (Class III or better)

- One ventilation fan (see appendix III for tool/kit descriptions)

- One circular saw (16-in., with $2\frac{1}{2}$ gal fuel)

- Two circular saw blades (16-in., diamond, continuous rim)

- Two circular saw blades (16-in., carbide tip)

- One pressurized water spray can

- Six canister-type respirators

- Twenty-four replacement canisters for respirators

- One generator (5-kW)

- Four floodlights (500-W)

- Six extension cords (50-ft)

- One junction box (4-outlet with GFI)

- One Wye electrical adapter

- One rotary hammer ($1\frac{1}{2}$-in.)

- One rotary hammer bit kit (see appendix III for tool/kit descriptions)

- One sawzall

- Twelve sawzall blades (wood)

- Eighteen sawzall blades (metal)

- One drill ($\frac{1}{2}$-in., variable speed)

- One drill bit set (steel, $\frac{1}{8}$-in. to $\frac{5}{8}$-in.)

- One drill bit set (carbide tip, $\frac{1}{4}$-in. to $\frac{5}{8}$-in.)

- One chainsaw (12-in. electric, with spare carbide tip chain if not already present from light inventory)

- One rebar cutter (1-in. capacity)

- One cutting torch (see appendix III for tool/kit descriptions)

- One Come-Along (2/4 ton)

- One demolition hammer, small (see appendix III for tool/kit descriptions)

- One demolition hammer, large (see appendix III for tool/kit descriptions)

- One extrication stretcher for confined areas

- Two shovels (folding, short)

- One mechanical axe (high-voltage)

- One mechanical grabber (high-voltage)

- Two pair lineman gloves (high-voltage)

- One upgrade high-pressure air bag to a total of 245 tons

- One air bag regulator control valve with 2 additional hoses

- Two building marking kits (see appendix III for tool/kit descriptions)

- One cribbing and wedge kit (see appendix III for tool/kit descriptions)

- One ram set powder-actuated nail gun (with 150 red charges)

- One box ram set nails with washers ($2\frac{1}{2}$-in.)

- One box ram set nails with washers ($3\frac{1}{2}$-in.)

- One green stone wheel (to sharpen carbide tips on tools)

- One set of nails (see appendix III for tool/kit descriptions)

- Two tri or speed squares

- Two framing hammers (24-oz)

- Two carpenter belts

- One level (6-in.)

- One level (4-ft)

- One nail gun, pneumatic (framing type, 6-p to 16-p)

- One case nail gun nails (8-p)

- One case nail gun nails (16-p)

- Thirty-two Ellis clamps

- One Ellis jack

- Eight post screw jacks

- Twelve screw jacks, pairs ($1\frac{1}{2}$-in.)

- Twelve pipes (6-ft x 1$\frac{1}{2}$-in., Schedule 40)

- Twelve steel pickets (1-in. x 4-ft)

- One case orange spray paint (line marking, downward application type)

- One case duct tape

- One technical search device (see appendix III for tool/kit descriptions)

- One hydraulic rescue tool (see appendix III for tool/kit descriptions)

Appendix III: USAR Tool and Kit Information Sheet

The following sheet provides additional information about the tools, equipment, and kits found in this chapter:

- Anchor kit

  - 1 box $\frac{3}{8}$-in. x 5-in. Hilti Kwick Bolt concrete anchors

  - 25 $\frac{3}{8}$-in. SMC stainless steel anchor plates

  - 25 $\frac{3}{8}$-in. Drop forged H/D eye nuts

  - Anchors and plates are for rope system anchor points

- Building marking kit

  - 2 Orange spray paints, line marking (downward) application type

  - 4 Lumber chalks

  - 2 Lumber crayons (red)

  - 2 Lumber crayons (yellow)

  - 4 Lumber pencils

- Camming device—Prusik loop (7-mm or 8-mm) or Gibb's ascender or combination of each

- Chainsaw (gasoline or electric) with carbide tip chain and one spare chain and bar oil

  - Gasoline—2$\frac{1}{2}$ gal spare fuel and oil mixture

  - Electric—need electric power source and 100-ft extension cord

- Chain set—One 1-ft with a grab hook on each end, one 5-ft with a grab hook and a slip hook, one 10-ft with a grab hook and a slip hook, one 20-ft with a grab hook and a slip hook

- Chain—All chain is $\frac{3}{8}$-in., grade 7 or better

- Cribbing and wedge kit

  - 24 Each 4-in. x 4-in. x 18-in.

  - 24 Each 2-in. x 4-in. x 18-in.

  - 12 Pair 4-in. x 4-in. x 18-in. wedges

  - 12 Pair 2-in. x 4-in. x 12-in. wedges

  - Containers to store and carry

- Cutting torch

  - One or more plasma cutter, exothermic torch with 50 rods

  - Heavy-duty oxy/acetylene torch with spare O2 cylinder or other similar device

- Demolition hammer (large) electric, pneumatic, or gasoline

  - 60-lb Minimum

  - 2 Each bull-point bits

  - 2 Each chisel-point bits

- Demolition hammer (small) electric, pneumatic, or gasoline

  - 30- to 45-lb Minimum

  - 2 Each bull-point bits

  - 2 Each chisel-point bits

- Edge protection—Commercial edge rollers, canvas tarps, split fire hose, or any combination of each

- Electrical detection device—Hot stick electrical alert device, volt/ohm meter, or other device to alert crewmembers of electrical current

- First-aid kit—Basic first-aid supplies for minor injuries to six victims or crewmembers. Examples of items to carry include:

  - Band-aids

  - Eye wash

  - 4-in. x 4-in. Gauze pads

  - Gauze dressings

  - Triangular bandages

  - Elastic bandages

- Friction device—Figure-eight with ears or brake bar rack or one of each

- Hydraulic rescue tool—Gasoline, electric, or manual device with 10,000-lb minimum force. Able to cut, spread, and pull.

  - Gasoline—$2^1/_2$ gal Spare fuel and oil

- Litter and complete pre-rig litter—Capable and rated for horizontal and vertical lift and hoist. Pre-rig can be commercial or preassembled to include adjustment and attachment capability.

- Nails

  - 25 lb, 16-p Vinyl-coated (green sinkers)

  - 25 lb, 8-p Vinyl-coated (green sinkers)

  - 25 lb, 16-p Duplex

  - High-humidity areas may require cadmium coated nails to prevent rust during long-term storage

- Pick-off strap webbing strap with one D-ring at one end and one V-ring adjuster on webbing strap

- Webbing—$1^3/_4$-in. wide with 10,000-lb rating, minimum 42-in. long

  - Hardware strength—5,000-lb Rating

- Rotary hammer bit kit

  - 1 Each carbide tip bits: $3/_8$-in., $1/_2$-in., $3/_4$-in., 1-in., $1^1/_2$-in., 2-in.

  - 2 Each bull-point bits

  - Appropriate adapters for bits and depth range capability

- Technical search device—One or more of the following: optical instruments (search cameras) and seismic/acoustic instruments (listening devices)

- Tool kit

  - 1 Each 12-in. crescent wrench

  - 1 Each 8-in. crescent wrench

  - 1 Each slip joint pliers

  - 1 Each channel lock pliers

  - 1 Each wire side cutters

  - 1 Each $1/_2$-in. socket set with ratchet and 6-in. extension

  - 1 Each $1/_2$-in. breaker bar

  - 1 Each ball peen hammer

  - 1 Set standard-head screwdrivers

  - 1 Set Phillips-head screwdrivers

  - Any other tools required for maintenance and repair of equipment in cache

- Trauma kit

  - Basic supplies to treat trauma injuries to six victims or crewmembers

  - ALS-type equipment (i.e. IV solutions, drugs, etc.) is not listed but may be carried if authorized

  - Examples of items to carry include: Large trauma dressings, splints, airways, bag valve respirator with large and small masks, etc.

- Ventilation fan—Electric or gasoline powered with extension tube to direct airflow

- Webbing Kit

  - 6 Each 1-in. x 5-ft

  - 6 Each 1-in. x 12-ft

  - 6 Each 1-in. x 15-ft

  - 6 Each 1-in. x 20 –ft

  - All webbing is spiral-weave nylon, 4,000-lb minimum-tensile strength. Each webbing length must be a different color

Endnotes

1 The LACoFD recently took delivery of two Firehawk helicopters as part of its effort to upgrade its fleet to exclusively twin-engine helicopters with heavier lift and better performance capabilities for firefighting, EMS, and technical rescue operations.

2 As of this writing, 8 of the nation's 28 FEMA USAR task forces are based in California, a factor that's related to the history of devastating earthquakes, floods, and other rescue-related disasters in the state.

3 Each FEMA USAR task force based in California is mandated by the state to have the capability to deploy at least one 16-rescuer component known as a swift-water rescue task force, for response to flood disasters in the state of California and other states that request river and flood rescue assistance. The state purchased the equipment and trailers for each of the swift-water rescue task force components, and each sponsoring agency is required to maintain the training and skills of its California-based FEMA USAR task force members. California-based swift-water rescue task forces have been successfully deployed and utilized to augment local river and flood rescue resources at a number of state flood disasters. As of this writing, there is no national requirement for FEMA USAR task forces to deploy for flood rescue disasters.

4 LACoFD Standard Equipment Inventory for USAR Companies.

5 There is a natural tendency to stick with what has worked in the past with regard to equipment, apparatus, and methodology. If a certain piece of equipment has been demonstrated to be effective under demanding field conditions, there is a tendency to purchase the same brand, make, and model (with appropriate upgrades) time after time. Some rescues simply prefer to use the tools that are most familiar to them. In many cases this proves to be the best approach. That said, it's still important for fire/rescue agencies to keep their options open and to conduct appropriate research and development that ensures their familiarity with emerging tools and technology that might demonstrably improve rescue safety, efficiency, and cost-effectiveness.

6 LACoFD Standard Equipment Inventory for USAR Companies.

7 Some equipment listed here is intended for the rescue of large and small animals. This is related to the local hazards and demands. In recent years, LACoFD engine and truck companies, USAR units, and helicopters have been called up to rescue horses, dogs, and other domesticated animals that have managed to become trapped (sometimes through the actions or negligence of their human friends) on cliffs, in canyon bottoms, in quicksand, on mud flats, in floods, or in overturned livestock trailers. Animals have also been found in sewers and storm drains, in water-filled wells, in deep well casings, in collapsed buildings, and a practically endless variety of other predicaments. The fire department has been called upon to help rescue the occasional bear trapped in a sump hole, deer entangled in fencing or trapped in mud, pythons beneath homes, and other unusual wild animal entrapments. The L.A. County Department of Animal Control, working in concert with the LACoFD, has developed animal rescue teams that respond with the fire department to provide expert guidance about the assessment, control, and treatment of animals during these rescue operations. See Volume III for details on animal rescues.

5

Rescue Training

Few topics are more essential to effective and reasonably safe rescue operations than an intelligent and rational regimen of training, continuing education, and exercises. Training for rescue is a topic that could fill an entire book. Realistic training is critical to prepare firefighters and rescuers to locate and extract trapped or missing victims without causing a secondary disaster (e.g. secondary collapse, rope system failure, swiftwater rescuers being swept away, etc.) and possibly suffering serious injury or death themselves. Effective training combined with experience during actual emergencies (and augmented by realistic simulations to develop recognition-primed decision-making skills) are among the most important factors in successful rescue operations. Certification in rescue-related knowledge, skills, and abilities (KSAs) is also an important goal of training because it provides verification that rescuers are actually qualified and ready for the task of rescue.

Increasingly, standardized certification is becoming a fact of life for rescuers who once got by with differing levels of KSAs that were often regional in nature. As the disciplines that make up *rescue* (and the systems that support them) become more organized and standardized, there is more emphasis on ensuring that rescuers are using similar methods, tools, and strategies. This can be both an improvement and a curse: standardization can have the positive effect of defining minimum levels of competency and allowing for a gradual development and spread of new methods, tool, and strategies. But standardization can also be a curse if it results in an overly dogmatic approach to rescue and if it is allowed to become overly bureaucratic, which can stifle the development of bold new initiatives. It's a mistake to allow those with less imagination to impose rules requiring that rescue be done only in one certain way, every time, all the time, all in the name of *standardization*. Most of the revolutionary improvements in rescue have come about because pioneers had the audacity to invent, propose, develop, and foster to completion a whole new set of methods, tools, disciplines, and systems that didn't previously exist. We don't want to lose that pioneering spirit in order to standardize rescue at the expense of innovation.

Here we should emphasize another important point about training: While it is critical for a successful approach to rescue and other emergency operations, training can never be a substitute for *actual emergency experience*. We all know that primary training and continuing education are the building blocks that prepare rescuers to decide and employ the most appropriate methods and techniques when confronted by an emergency. But we also know that top notch rescuers stretch the limits by going far beyond the basic buildings blocks. The most effective rescuers include in their preparedness regimen factors like research and development in new knowledge, methods, and equipment to do the job better, faster, and safer. They also include experience at the scene of actual emergencies, where they actually conduct, supervise, or observe technical search and rescue operations and other emergencies. This helps them by building libraries of "slide programs" in their heads, which in turn help them quickly react to unusual and fast-changing conditions in future emergencies. In other words, training should not be considered the end goal, but a stepping stone toward more effective emergency operations.

The tendency of some trainers and instructors to place *training* on a higher plane of importance than *emergency experience* is one that should be countered by a rational understanding that *actually and successfully doing the job* is every bit as valid (or even more valid) than *practicing to do the job*. Although we want to emphasize, expand, and support better training, we should not allow a "bean counter" mentality to pervade the world of rescue by substituting knowledgeable and highly experienced trainers with clipboard-carrying bureaucrats whose main goal is to "check off the boxes" on evaluation sheets during rescue training.

We want to foster innovation, experiment, and independent thinking while we teach standardized methods of rescue. And we want to avoid the situation where highly experienced rescuers are prevented from doing their jobs because they are using their common sense and innovative skills to develop and practice new ways of doing things that may go against the grain of the "standardized training regimen." This will have the effect of inhibiting innovation, experimentation, and finding better ways to locate and rescue victims that has been the defining hallmark of modern rescue improvements.

At some point, as our systems of rescue response and rescue training become more organized (and sometimes more bureaucratic), we need to ask ourselves the two most basic questions about rescue skills, methods, and strategies:

1. Will it work?

2. Is it safe?

In a perfect world, good primary training, supported by research and development and continuing education, would be further augmented by vast quantities of actual emergency experience. For personnel with responsibility for search and rescue, this means responding to as many search and rescue emergencies as possible. How do fire/rescue agencies accomplish this goal? For starters, they can ensure that specialized rescue units are dispatched to all technical rescues that are feasible, even if the need for their services are not absolutely certain.

The good that will be gained by personnel attending large numbers of technical rescues and other related emergencies cannot be overemphasized. Some agencies have adopted the practice of responding with one or more additional rescue/USAR units to the scene of significant rescue operations in order to assign them as rapid intervention teams *and* to gain more experience by assisting, supporting, supervising, or observing the rescue. This provides personnel with additional opportunities to assess rescue conditions, identify hazards, develop strategies (and alternate plans), observe the use of equipment to determine its effectiveness for certain conditions, and determine how certain strategies either do or do not work.

In some departments, it's become accepted practice for rescue/USAR units to continue responding (usually non code-r) to the scene even after the IC has determined their services aren't needed. This helps them to observe the conditions that caused the rescue emergencies, to observe what was done to mitigate hazards and rescue victims, to act as safety officers and/or rapid intervention teams, to help critique the incident after it's over, and to help them gain a better understanding of how they would handle the same situation if it happens again. In some departments, the IC is encouraged to keep the rescue/USAR companies coming even after the rescue has been assured, in order to allow the rescue personnel to observe the situation and to support the concept that says the more experience they have under their belt (and the more rescue operations and conditions they observe), the better prepared they will be for future rescues.

When it comes time to assess the KSAs of rescuers, their performance and success conducting or supervising actual rescue emergencies should be taken into account. Some training and recertification systems rely exclusively on KSA testing conducted by trainers with clipboards and check-off sheets to review what is sometimes remedial training (things already taught and learned in primary training). But more progressive systems incorporate the experience, knowledge, lessons, and performance achieved by rescuers in the course of handling actual rescue emergencies. They also take into account the experiences of rescuers from other agencies and other regions that have confronted challenging or unusual rescue emergencies.

Other progressive systems incorporate realistic simulations and place rescuers in positions where they are expected to perform at a level higher than that taught in basic training. For example, rescue officers should go beyond the basic manipulative rescue skills they learned in primary training, and they should practice and be evaluated in the KSAs that are more pertinent to their position. (This may include providing technical advice to the IC, supervising a rescue group, a search group, a division; directly supervising rescue tasks performed by their crew; performing rescue tasks in support of their crew; or in some cases, conducting the actual rescue themselves). In a world with ever-expanding rescue hazards and challenges, this represents a balanced approach to rescue training, certification, and continuing education.

Another important factor in rescue training is the use of case studies and researching fire/rescue periodicals and other available literature about rescue emergencies and operations. Today there are a number of magazines that specialize in fire and rescue preparedness and operations. They include case studies of past emergencies and how they were managed (as well as discussions about what could be done to improve future operations); articles on cutting edge equipment and methods; discussion about emerging technology, tactics, and strategies; forums to discuss and debate current issues and controversies; letters from readers to inquire or correct; and other valuable information. Naturally, textbooks are another tool in research, development, and preparedness. The use of training videotapes, compact disks and DVDs, and cable systems to transmit important lessons about new rescue methods and equipment and to conduct case studies are important new factors in training. In Figure 5–1, teams practice rappelling from helicopters onto the roof of high-rise buildings on a tower designed to support this "ground" training. This happens prior to rappelling from a hovering copter and for periodic skills review without the necessity of flying the copters for training.

Fig. 5–1 L.A. County FD USAR Company Personnel, Assigned the Role of Helo High-rise, Practicing Rappelling with Full PPE from a Specialized Training Tower

There are many other options for training firefighters in the basics of conducting rescue operations in a professional manner with a reasonable level of safety, too many options to list all of them here. There are a growing number of advanced rescue courses intended to take those who have completed basic training and raise their level of skill and knowledge, including those intended to meet the requirements of NFPA 1670 and other applicable standards. The number of formal training options is expanding every year, enabling fire departments to tailor their rescue training programs to the needs of the local agencies. The fact that we can't list all the current training courses and options is a very good sign indeed; 20 years ago it would have been quite easy because the selection of formalized courses was relatively sparse.

Training as Risk Management

Here is a question for the ages: Why do we train? Risk management expert Gordon Graham argues that there are essentially two types of emergencies, both of which we as firefighters and rescuers must be trained to handle: *low-risk, high-frequency* emergencies, and *high-risk, low-frequency* emergencies.

It's easy to see why it's simpler to be prepared for low-risk, high-frequency incidents. Once the initial training is completed, firefighters and rescuers usually get a chance to practice these operations repeatedly, simply because they are high-frequency emergencies. And because they are also low-risk operations, the consequences of mistakes are less likely to cause serious injury or deaths to firefighters and rescuers.

An example of a low-risk, high-frequency operation is an EMS response to a single-family residence in the typical community. In the course of a typical fire service career in a typical fire department, the average firefighter may spend many hours performing patient assessments, getting vital signs, obtaining medical histories, treating patients with oxygen, and other forms of EMS assistance in single-family dwellings. These are low-risk emergencies in the sense that few firefighters or rescuers become lost, injured, or trapped during the course of performing EMS operations in single-family homes. And the sheer number of EMS emergencies to which the fire service responds tells us that they are extremely high-frequency operations. So, while firefighters definitely require proper EMS training to perform this function, the typical firefighter responds to actual EMS calls practically every day (in some cases up to a dozen or more times a day) and becomes well-practiced in dealing with the idiosyncrasies and relative hazards of these operations.

Much the same can be said for other operations that are typical for local firefighters and rescuers. Whereas the typical firefighter may be well versed in EMS work, he or she may have no experience being lowered into a deep-shaft confined space to rescue an unconscious victim who's physically trapped (a low-frequency/high-risk incident). Firefighters working in jurisdictions where tenement fires are typical tend to be well versed in putting out fires in tenements. Those working where wildland fires are common become expert at managing wildfires. Fire stations in places where the population is predominantly elderly are accustomed to handling heart attacks, CVAs and related EMS responses, and so on. In short, we become very good at the things we do all the time. A facility designed for training is shown in Figure 5–2.

Fig. 5–2 USAR Firefighters Practice Tunneling and Breaching Skills for Structure Collapse Operations

Graham points out that the truly difficult emergencies tend to be those categorized as high-risk/low-frequency. Low frequency implies that the incidents are relatively rare, which breeds unfamiliarity and works against automatic responses by personnel. High risk implies that firefighters and rescuers may be seriously injured or killed if something goes wrong and that the potential for things going wrong is relatively high. One can argue that training becomes more critical for these events because their unusual nature makes them unfamiliar to us and because they are more likely to kill or maim rescuers if they don't take the correct actions without delay (Fig. 5–3).

Fig. 5–3 Confined Space Exercises Challenge Rescuers to Try Innovative Approaches

Effective training is one of the only hedges we have against a lack of experience conducting low-frequency/high-risk operations. The good thing is that effective training also helps firefighters operate safely and effectively in high-frequency/low-risk emergencies by correcting bad habits and introducing new information about familiar (high-frequency) hazards.

Avoiding the blank screen syndrome through RPD

Preparing firefighters and rescuers to manage high-risk/low-frequency emergencies involves ingraining their minds with workable solutions that they may draw from in the midst of urgent situations where time is critical. In these situations, firefighters and rescuers don't have all the information they need to make the right decisions, and one mistake can prove catastrophic. This dilemma was highlighted in a study by Chief Deputy Larry Miller (retired), former Director of Training for the LACoFD.[1]

According to Miller, when firefighters and rescuers are confronted by incidents that they have not yet experienced in training, continuing education, simulations, or previous emergencies, they are forced to operate without a mental *prototype*, or mental slide program of possible solutions based on past experience or study. In this view, the mental prototype is a sort of guide that helps firefighters and rescuers select the most appropriate strategy and tactics to resolve a time-critical problem during the course of an emergency (Fig. 5–4). In the following photo, firefighter/paramedics practice high angle "pick-off" operations at a training facility developed for helicopter-based rescue operations.

Fig. 5–4 L.A. County Fire Department Air Operations Personnel Practicing Cliff Rescue "Pick-offs"

During his research on the matter, Miller found that in the 1980s the U.S. Army undertook a study of the process known as rapid decision-making (RDM) under emergency conditions. The purpose of the Army study was to better learn how people make time-critical decisions under fire without all the data normally required to select the most appropriate strategy and tactics. The results of the Army study, in turn, were to be used to develop more effective training and preparation for military officers, who are expected to make time-critical decisions (often with limited battle experience). The theory was that it might be possible for the military to compensate for the relative lack of actual battle experience among many of its officers by learning more about the process of RDM and applying the concepts to its officer training programs.

To conduct the study, the Army looked at all the civilian professions, trying to determine which group was most frequently confronted with the kinds of decisions faced by military officers in battle. The group selected would be making decisions that were time-critical ones in which they would lack all of the information normally required to make a decision. Their decisions would be made in a setting of extreme physical danger with the lives of their colleagues, allies, and citizens hanging in the balance. The Army settled on civilian fire department officers because they are most commonly called upon to conduct RDM under conditions of personal and shared physical danger in the course of their daily work.

A test group of professional fire officers was selected. This test group was followed as they prepared for and responded to fire and rescue emergencies and performed RDM. Ironically, most of the fire officers didn't even know what RDM was. Like most fire/rescue decision-makers, they had been doing it for years without ever naming the process.

The Army researchers found that many of the time-pressure problems experienced by the fire officers reduced their ability to make decisions consistent with standard operational guidelines. Instead, firefighters and their commanders tend to fall back on intuition, gut feeling, and natural reaction when faced with immediate danger situations that require instant solutions. In other words, fire department officers forced to make time-critical emergency decisions appeared to strongly rely on mental models (or mental slide programs) of similar situations that they had stored in their heads (Fig. 5–5). Repetitive practice of high-risk operations like confined space rescue helps prepare "slide pro-grams" in the minds of rescuers and their commanders who will be expected to make instant decisions during the course of actual emergencies where lives are on the line.

Fig. 5–5 Confined Space Rescue

When confronted by situations in which they didn't know all the facts, yet which demanded immediate deci-sions to initiate actions to save lives and property and to protect the lives of firefighters in the process, the fire department officers appeared to mentally flash back. The officers would recall their own mental slides of similar situations, then choose those that in the past had led to positive outcomes.

In other words, when there was no time to thor-oughly analyze all the facts, and when a critical decision was needed *now,* they didn't just *guess.* On the contrary, the officers made rapid mental calculations based on a mental image (or flashback) of similar situ-ations they had experienced either in training or previous emergencies. They quickly compared the conditions at hand against images in their personal flashbacks, matched the mental slide that most closely approximated the conditions on the scene and had the most positive outcome, and chose that action. This process allowed the most experienced and most effec-tively trained officers to make rapid and appropriate decisions that led to positive outcomes. For officers who found that they had no such mental images because they had yet to experience similar situ-ations under actual emergency conditions, decisions came slower and were more likely to be flawed. Those who had read about similar situations or experienced them in training simulations seemed to have an advantage over those who had no preparation at all.

This is an important point about training. When fire-fighters or rescuers have one or more accurate slides in their heads of the type of incident they are facing, they may be better able to visualize how and why the emergency is happening, what is likely to happen next, and how similar situations were successfully resolved in the past. This mental model is referred to by the Army as a *prototype.* The prototype doesn't necessarily have to sprout from an actual emergency experience. The act of reading about it, thinking about how to handle it, or simulating it appears to leave a prototype imbedded in peoples' minds. It's perhaps the next best thing to experiencing the actual emergency (Fig. 5–6).

Fig. 5–6 LACoFD USAR Company Members and First Responders Working to "Corral" and Safely Extract a Psychotic Man Threatening to Jump from a Railroad Trestle.

This incident required eight hours to complete because the man was found running at full speed across the entire width and breadth of the trestle framework, and his paranoid condition compelled him to avoid rescuers who were attempting to help him. With the public and news media watching and filming every more, and with law enforcement unable to immobilize the man with more traditional non-lethal methods for fear of causing a fatal fall, this rescue required the combined efforts of the fire department and law enforcement (with a crisis negotiations team). In the end, it took the coordinated operations of 2 aerial ladders, 1 ladder tower, 1 USAR company applying high angle methods, an L.A. City Fire Department truck company with fall bags, and 2 crisis negotiators to safely extract the man. This is an example of emergencies for which there are no "play books," but which can be managed by applying the "slide program" that most closely matches the conditions.

The Army found that with proper input (e.g. emergency experience, study, simulation), the decision-maker's mind can be primed with prototypes that help them accurately recognize current conditions and future probabilities. Almost subconsciously, the person matches the current actual conditions with the mental prototypes in order to make rapid and effective decisions. This is what the Army calls a recognition-primed decision (RPD).

We can infer that it's especially important to prime the minds of firefighters and rescuers with accurate prototypes to enable them to resolve the problems that accompany many rescue operations, especially those that represent high-risk/low-frequency emergencies. We can do that through actual emergency experiences, through case studies, through discussions in training settings, and through rescue simulations. This represents an opportunity to improve the decision-making ability of firefighters, rescuers, and officers when we can't repeatedly expose them to every type of actual rescue emergency. In other words, through effective training, we can implant accurate and helpful prototypes in their minds that will help them make the best decisions under fire.

Training the Computer Under the Helmet for Optimum Performance

Realistic training is a proven path to safer and more effective rescue operations. In (Figs. 5–7 and 5–8), firefighters practice confined-space rescue in a railcar donated by a railroad that also installed tracks at a regional fire department training center.

Fig. 5–7 Confined-space Rescue Training

Fig. 5–8 Firefighters Practice Confined-space Rescue in a Railcar

The Army's RPD study brings to mind a favorite concept of Ray Downey, longtime captain of FDNY Rescue 2 and one of the most experienced rescuers in the nation before his death in the 9-11 attacks. Downey liked to remind firefighters and rescuers that charts, manuals, field operations guides, pneumonics, and electronic organizers are fine as aids for the decision-making process at the scene of a rescue or fire, but the most important tool a firefighter or rescuer possesses is the computer under his helmet.

Downey believed that if a firefighter/rescuer's mind is programmed with the right information, they will be prepared to quickly scroll through the potential options (prototypes) and select the best one—in an instant. Downey's approach, stated in a different way and apparently developed long before the aforementioned Army study, closely paralleled that of RPD (Fig. 5–9).

Fig. 5–9 Extracting Victims of a Confined-space Rescue Situation

Downey believed that if (or when) conditions change unexpectedly during the course of rescue and fire-ground operations, the firefighter's computer under his helmet is capable of recognizing the problem and calculating the solution much faster than any machine ever devised by man. A key to this, Downey believed, is to ensure that the computers under our helmets are programmed with the most accurate data possible through good training, physical experiences in actual incidents or simulations, mental experiences from emergency exercises, case studies, and the like.

What is the relationship between RPD, the computer under the helmet, and rescue training? The answer seems to be that in the absence of experience managing the effects of actual rescue emergencies, the next best thing is to simulate those experiences through solid primary training, effective continuing education, and realistic simulations of rescue emergencies.

It's rare to find a single fire station or rescue unit that responds to every kind of rescue emergency each year to provide their personnel with experiences that they could translate into RPD. That's why it's so important for fire/rescue agencies to support and provide realistic simulations and continuing education for firefighters and other rescuers whose computers under their helmets are in need of mental slide shows that will enable them to make the best decisions under unusual and high-risk circumstances.

Who Are the Trainers and What Are They Teaching?

There is an old saying that those who can do, do, and those who can't do become trainers. That may have been true at one time, but with regard to rescue, it's time to lay that saying to rest and bury it. While it's true that there will always be a need for experienced and accomplished retirees and those injured in the line of duty to teach, it is more important than ever to ensure that you have a cadre of instructors who are active-duty members who actually conduct rescue operations.

It's important to ensure that the information being taught is up-to-date, that it's actually being used in the field, and that it actually works under emergency conditions. It's more important than ever to review case studies to highlight successes (what worked) as well as failures (what didn't work). It's necessary to examine how things are done in different agencies across the nation (and, yes around the world) to select the most practical and successful training points about tactics, strategy, methods, etc. Sometimes the best way to do that is to use trainers who are actively engaged in managing the emergencies, as well as those with the wherewithal to conduct research and development.

Clearly, it's also important to select instructors who know how to teach. We all can remember instructors who might have been brilliant tacticians or who clearly excelled in doing a job but who simply were terrible instructors because they had not been instructed in the methodology of teaching. This is one reason why some local, regional, and state fire training academies and colleges require instructors to complete teaching courses like Fire Instructor 1-A and Fire Instructor 1-B prior to being certified to instruct these topics. It's a reason many fire departments require their fireground and USAR instructors to complete fire instructor courses prior to teaching. It's why FEMA requires its USAR course instructors to complete instructor training specific to the discipline being taught. This is why the National Fire Academy and other major fire service organizations emphasize good teaching skills as well as experience and subject matter knowledge of fire service instructors.

Physical hazards associated with rescue training

Naturally, effective rescue training and simulations have some inherent risks that mirror those found in emergency operations. In fact, it can be stated that the best rescue training includes realistic simulations that mirror some of the real-life (potentially injurious or even lethal) conditions found on the rescue scene. The difference is that under actual rescue conditions, there is always a level of uncertainty about changing conditions and hazards. In training however we take whatever precautions are necessary, including the use of belays, backup air supplies, and RICs, etc. to ensure that personnel are not seriously injured or killed during training sessions (Fig. 5–10).

Fig. 5–10 Rescue Technicians from Active Rescue Units as Instructors—A Good Concept

How can training hazards be reduced to a reasonable level? One method increases the use of didactic (classroom) training to accomplish the following:

- Communicate policies and procedures

- Conduct preplanning and blackboard simulations

- Critique past incidents

Nevertheless, classroom sessions are no substitute for effective manipulative training under realistic rescue conditions. For first responders and rescue technicians alike, realistic manipulative training is a key to safe and effective operations. For that reason, it's important for supervisors to embrace rescue training and encourage firefighters to practice as if their lives depend on it. In fact, survival often does depend on instant reactions to changing conditions, and the person you rescue may be your own partner or teammate.

In the United States, NFPA 1670, OSHA regulations, and other recognized standards form the basis for determining what training is required to accomplish the desired level of readiness to manage local, regional, state, and national hazards. Each fire/rescue agency should correlate its training standards with the local and regional hazards it faces.

State fire marshals, the National Fire Academy, FEMA, NFPA, the National Association for Search and Rescue (NASAR), the IAFF, the IAFD, and other organizations have taken lead roles in developing, adopting, recommending, and requiring ever more stringent, quantifiable, and effective rescue training standards. These are positive trends that support the efforts of local fire/rescue agencies to adopt effective and appropriate training for the rescue hazards likely to confront their personnel (Fig. 5–11).

Fig. 5–11 Confined-space Rescue Training Props Allow Side-by-side Comparisons and Sharing of Lessons

Example of helicopter rescue training

To illustrate some of the concepts of high-risk rescue training as discussed in this chapter, it's instructive to review how these concepts are actually employed in various fire/rescue agencies. Helicopter rescue training is a good example of a high-risk learning environment in which the use of standardized curriculum, minimum-performance standards, and qualified/highly-experienced instructors minimize the risk of training danger. Helicopter rescue training minimizes lethal training dangers through intelligent instruction and practice and realistic training to support the process of RPD. Because the helicopter environment is quite unforgiving, it's a good example of how high-risk rescue training can be

successful if certain precautions are taken and innovative approaches are considered.

To reduce the relative risks associated with helicopter rescue operations, most fire/rescue agencies with helicopter fleets devote significant resources to ensure the best equipment, the best mechanics and pilots, and the most qualified crewpersons. Not surprisingly, training has long been identified as a critical factor in the equation of safe helicopter rescue operations (Fig. 5–12).

Fig. 5–12 Some Fire Departments have Entered into Partnerships with Corporate Neighbors to Help Distribute the Cost of Developing and Installing Training Props

The process of training within the helicopter environment has inherent risks that mirror many of those found during actual emergency operations. Naturally, preparing fire/rescue personnel to conduct hoist rescue, short-haul rescue, helo-swift-water rescue, helo-high-rise firefighting and rescue, and other helicopter-based evolutions requires the use of helicopters in flight, at least in part.

While classroom sessions and research are required aspects of the training regimen, helicopter rescue training eventually requires helicopters to hover with personnel standing on the skids, in the cabin, and suspended below the copter. To practice various evolutions, rescuers must become accustomed to the noise, the rotor wash, and the aerodynamics of dangling below hovering helicopters. They must practice reacting to the changing conditions that will affect their operations. It's necessary to practice these operations in differing conditions of light, weather, and other factors that will affect emergency operations.

Training in these realistic conditions ingrains in the rescuers the ability to recognize and assess the conditions, anticipate changing conditions, and make rapid decisions based on the prototypes they have developed. This is part of the process of improving RPD. It often involves assigning personnel to act as simulated victims on the ground, on cliffs, on buildings, or in the water to simulate actual rescue scenarios. In cases where using live simulated victims is unsafe, mannequins and other rescue aids may be employed, but rescuers must still operate in an unforgiving environment where there are real consequences for certain critical mistakes. That is why safety is always a primary concern during training of this nature.

Even while training, hovering helicopters can be unforgiving. There is ever-present potential for flight mishaps, including rotor strikes, engine failures, and engine fires. Human error can cause collisions with objects. Rescuers dangling on cable or short-haul lines can collide with trees or rocks. Rescuers can also fall or be struck by items like caribiners falling from the helicopter.

Therefore, helicopter training requires some of the same precautions that we use during actual rescues, sometimes even more stringent precautions. If we can avoid the risk of working in or below an actual hovering helicopter to repeatedly practice skills such as rappelling from a copter or being hoisted from a copter, it will be an advantage. If we can do it using a reasonable facsimile of a hovering helicopter, without the lethal risk of a hovering copter, the exercise will be that much better.

Also of concern (although always secondary to personnel safety) is the matter of cost. There is a significant hourly cost (including fuel and maintenance) associated with helicopter flight, even if it is for emergency duty or training. Therefore it simply makes economic sense to limit helicopter flight time where it is feasible to do so without jeopardizing the mission of the helicopter rescue team.

In what ways can helicopter hoist, high-rise, swiftwater, and short-haul training hazards and costs be reduced to a reasonable level? Use didactic training to communicate policies and procedures, conduct preplanning and blackboard simulations, and study past incidents.

Didactic training is effective and absolutely necessary in many regards, but classroom sessions are no substitute for effective manipulative training using actual helicopters in the air. You may ask if those helicopters always have to hover in the air under engine power for manipulative training to be effective. The answer is No. Fortunately, some fire departments and rescue agencies have discovered that a combination of in-flight training and mock-up helicopters mounted high above the ground provides an effective training regimen for helicopter rescue teams (Fig. 5–13).

Fig. 5–13 "Hot Flight" Training for USAR Company Personnel Assigned as Helo High-rise Team—Because of Inherent Risks with Helicopter Flight, Strict Safety Protocols are Needed

This is the problem that confronted the LACoFD, which operates a fleet of two Firehawk helicopters, four Bell 412 helicopters, and a Bell Long Ranger. The LACoFD helicopters conduct rescue operations daily in terrain that includes 10,000-ft high mountains, coastal cliffs and waters, vast urban areas, lakes and reservoirs, high-rise buildings, and other places where helicopters prove advantageous for rescue. Maintaining the readiness of the firefighters and pilots assigned to these helicopters requires an effective training program. One of the biggest challenges is maintaining the skills of the air operations and USAR task force firefighters in high-risk operations like hoist rescue, short-haul rescue, rappelling, and other related operations.

To accomplish the goal of effective, repetitive, and reasonably safe hoist and short-haul rescue training at reduced flight costs and risk, the LACoFD built a three-story tower upon which was mounted an actual helicopter (sans engines and transmission) with an operating hoist and other training aids. This concept is not new by any stretch. A number of civilian agencies and military units across the United States and other countries use similar aboveground helicopter mock-ups for training. But this one is somewhat unique in the range of rescue operations that it's used for and the manner in which it was procured and built.

In the early 1990s, a group of firefighter/ paramedics,[2] pilots, and mechanics assigned to LACoFD air operations, USAR, and swift-water programs proposed the development of a tower-mounted helicopter prop. This prop would enable rescuers to spend more time practicing manipulative skills without the need to burn up blade time. The concept went through several iterations until 1994 when the project was green-lighted to be built at the department's air operations facility.

One advantage of this project was its cost-effectiveness. The LACoFD acquired a surplus Bell 205 helicopter, which was stripped out by the air

operations mechanical staff and painted in the standard scheme for the department. Breeze Eastern donated and installed a brand new rescue hoist on the copter (Fig. 5–14).

Fig. 5–14 The Inherent Risks of Flight can be Avoided for Helicopter Rescue Training at Custom Facilities

The tower itself is actually part of a petroleum-cracking tower that was being disassembled and removed from Shell Oil's Carson refinery. Shell donated the tower for this project. The tower and its mountings were designed by LACoFD Battalion Chief Don Hull and engineered by a private engineer hired by the department. With the engineering completed and the building permit obtained, the foundation was laid with embedded tower connections, and the tower was later lifted into place with a 60-ton crane.

This training facility provides an effective means of conducting certain types of repetitive manipulative training sessions while reducing wear and tear on the department's helicopters. It also improves some training safety conditions by eliminating the need to have an actual helicopter hovering in the air at all times (Fig. 5–15).

Fig. 5–15 USAR Company Members Practicing Loading Procedures for Helo High-rise Operations, Prior to Rappelling out the Door on the Opposite Side of This 3-Story Training Prop

Fig. 5–16 LACoFD Air Ops Firefighter/Paramedics Practicing "Pick-off" Rescue Methods Using Tower-Mounted Copter

All air ops-qualified personnel and those permanently assigned to LACoFD USAR task forces are required to perform a hoist rescue or short-haul rescue during training or emergencies at least every 90 days to maintain their qualifications. Considering the number of personnel who require primary and continuing hoist rescue training and the number of hours required to support this training, the tower has dramatically reduced the flight hours required of the LACoFD helicopters (Fig. 5–16).

While this does not eliminate the need to conduct hot training and certification sessions with hovering helicopters, it does provide a means to simulate a variety of rescue conditions. These conditions include night operations, pick-off rescues from cliffs and burning buildings, and rescue litter operations. The tower enables these training operations while reducing wear and tear on key components and reducing some of the inherent training risks.

Appendix I: Examples of Rescue Training Standards

There are a number of applicable rescue training standards with more under development. Here is a sample of national rescue training standards in the United States:

- NFPA 1670, *Standard on Technical Rescue*

- NFPA 1006, *Standard on Professional Competencies for Technical Rescue*

- NFPA 1983, *Standard on Life Safety Rope, Harnesses, and Hardware*

- NFPA 1470, *Standard on Search and Rescue Training for Structural Collapse*

- NFPA 1500, *Standard of Fire Department Occupational Safety and Health Programs*

- OSHA 1910.134, *Respiratory Protection*

- OSHA 1926.650, *Trench and Excavations*

Endnotes

[1] Miller, Larry, Recognition Primed Decision-Making on the Fireground: How to *Avoid the Blank Screen Syndrome*, American Fire Journal, April 1996.

[2] One member of this group, firefighter/ paramedic Jeff Langley, a pioneer in the field of helicopter and swift-water rescue, died during the course of a helicopter hoist rescue operation in Malibu in 1993. The helicopter rescue training tower has since been dedicated and named in his honor.

6

Commanding Rescue Operations

Effective command of rescue operations does not require one to be an expert in technical rescue. However, it helps to have a good working knowledge of what needs to be done, who needs to do it, and how to support them. Not surprisingly, the same qualities that define a good fireground commander are effective when the emergency involves a technical rescue or a rescue-related disaster. Because technical rescues may tend to occur less frequently and require more unusual tactics, strategies, and resources, many ICs will not have experience managing the particular type of problem at the scene of a major rescue and therefore may place greater reliance on technical advisors assigned to rescue units and others with knowledge about the hazards they are facing. Some ICs may not be familiar with the resources required to manage highly technical "career rescues" (e.g., those that you may see only once in a career, if ever), or disasters that require the response of regional, state, or even federal rescue teams like FEMA USAR task forces. Recognizing these challenges, this chapter is intended as a primer in the management and command of technical rescue emergencies.

Responsibilities of the Rescue IC

ICs at the scene of rescue-related emergencies and disasters are responsible for the followings tasks:

- Initiating first responder operations within the parameters of the agency's rescue/USAR training and guidelines

- Recognizing if and when the safe working limits of first responders have been reached

- Ensuring effective actions are being taken by qualified personnel to secure and stabilize the scene until the arrival of local USAR/rescue units, helicopters, secondary responders, tertiary responders such as state and federal USAR task forces, and in some extreme cases, international USAR teams

- Utilizing these resources to their greatest advantage when they arrive

The rescue IC is responsible for requesting additional resources based on the need for any of the following:

- Specialized equipment and trained personnel to establish rigging, operate rope systems, and staff litter teams

- RICs, when needed

- Personnel for hauling systems and carrying patients out of back-country locations

- Helicopter transportation of rescue resources when appropriate to expedite resolution of the incident and to improve personnel safety

He should make decisions that comply with applicable worker safety regulations for operations in environments characterized as immediately dangerous to life and health (IDLH) and where two-in/two-out, rapid intervention, and personnel accountability protocols are required.

Unless there are compelling reasons to do otherwise, the IC should refrain from canceling rescue resources until one or more of the following benchmarks is achieved:

- Rescue of the victim(s) has been completed and rescuers have exited any IDLH environment or other special hazard situation

- Successful and safe rescue of the victim(s) has been positively assured (i.e. beyond a reasonable doubt) based on the eyewitness assessment of a qualified rescuer, and rescuers are no longer located in an IDLH environment

- A size-up indicates that the incident is strictly a first responder operation and does not require secondary responder capabilities, nor is there a need for secondary responders to perform the function of safety officer, structural safety assessment, RIC, or another assignment based on their specialized capabilities

- The incident has been confirmed to be a false alarm, i.e. there is no victim, or the victim has been located and determined to be in a location of safety with no need for technical operations to relocate, to receive EMS assistance, etc.

Responsibilities of Rescue/USAR Company Officers

Rescue/USAR company officers are responsible for ensuring constant readiness of their companies to support the first-arriving firefighters in order to effectively and safely manage the consequences of rescue emergencies anywhere within a fire department's jurisdiction. This includes planning, conducting, and participating in mission-specific training exercises. It also involves preplanning *high probability* and *extreme risk* rescue sites within a department's jurisdiction and coordinating with allied agencies to ensure maximum effectiveness. Company officers also maintain all pertinent rescue skills (Fig. 6–1).

Fig. 6–1 Who is in Charge of Your Rescue Scene?

In agencies that have access to helicopters, rescue/USAR company officers may be responsible for assessing the transportation needs for incidents in remote or inaccessible areas and requesting helicopter transportation for rescue company members and equipment when airborne response is likely to materially impact the outcome or personnel safety of the incident. They may also have the authority to order the rescue company to respond via helicopter for technical rescues where the rescue company's assistance may be required to conduct hoist rescues, short-haul rescues, or ground-based rescue in conjunction with the helicopter crew.

Key points on managing rescue emergencies

The majority of USAR-related incidents will be resolved by first responders. Recognizing this, it becomes evident that all first responders should be prepared to size-up and continually evaluate USAR-related emergencies in order to:

- Recognize critical cues and life hazards for victims as well as rescuers, including those that indicate the need for secondary responders to help manage them

- Initiate effective first responder operations within the scope of their training and equipment

- Recognize the boundary between the safe limits of first responder and secondary responder operations and understand when it's best to concentrate the efforts of first responders on stabilizing the situation as best they can until the arrival of secondary responders

- Request additional resources based on the need for specialized equipment and trained personnel to conduct stabilization, search, rescue, and rapid intervention operations

- Recognize when it's appropriate to expedite the response of USAR/rescue units and other resources by requesting or approving helicopter transportation

ICS/SEMS

Regardless of the strength and capabilities of the rescue resources, they cannot be used to their maximum effectiveness unless all responding agencies, firefighters, rescuers, and law enforcement personnel are using some form of the ICS or SEMS. This is especially true with regard to disasters, which by their very nature require extraordinarily strong command-and-control to ensure the most good is being done for the most people with resources that may be overtaxed or overwhelmed (Fig. 6–2).

Fig. 6–2 Members of FEMA USAR Incident Support Team Developing Strategy for Collapse SAR Operations at the Pentagon following the 9-11 Attacks

Unless some sort of ICS is being used as a baseline for command and control of emergency operations, rescue operations are going to be hampered. Some responders unfamiliar with incident command have mistakenly developed the impression that ICS/SEMS is some sort of *entity*—an actual object or organization—rather than a *tool*. About this we must be absolutely clear: ICS and SEMS are neither agencies, nor schools of thoughts, nor organizations. Rather, ICS and SEMS are generic emergency management tools designed for a single IC (in some cases, a *unified* command) that builds or reduces an IC organization based on the needs of the particular emergency incident being commanded.

ICS and SEMS enable the IC to delegate to designated officers the major functions of incident safety, operations, planning, logistics, and in very large or long-term incidents, administration/ finance. ICS/SEMS is consistent with the time-proven command-and-control concepts of manageable span of control (optimum and maximum number of persons that one supervisor can command with maximum effectiveness), delegation of authority, company unity, and unity of command.

Within the major function of the operations section, the IC of a major rescue emergency (through the operations section chief) can manage significant functions such as an SAR branch, a suppression branch (if needed), or a hazmat branch. Within the scope of authority of the branch directors would be teams that might include search groups, rescue groups, medical groups, hazmat groups, or air operations all possibly within staging areas or geographical divisions. In disasters of sufficient size/magnitude to warrant the deployment of FEMA USAR task forces, the IC could integrate the FEMA IST into a command structure to help manage SAR operations and/or the operations of the USAR task forces (Fig. 6–3).

Fig. 6–3 Company Officers Attending an Operational Briefing in Preparation for a Confined Space Entry

For smaller incidents, the functions of search, rescue, hazmat, EMS, air operations, and staging could be managed by the operations section chief. In many cases, daily rescue incidents are resolved with just three ICS positions: IC, rescue group leader, and medical group leader. If the incident also involves fireground or hazmat operations, the IC may assign an operations section chief to manage the rescue, suppression, hazmat, and medical groups.

It should be evident based on these brief examples that ICS/SEMS is highly flexible and expandable (or shrinkable), depending on the needs of the incident. It's popular to describe ICS/SEMA as a sort of toolbox that the IC opens to grab the tools he needs to handle the incident.

The simple fact is, large and/or complex emergency operations cannot be conducted with maximum efficiency and a reasonable level of safety for trapped victims and rescuers unless there is an effective, modular, expandable, and highly flexible command-and-control system in place *before* the emergency occurs. Therefore, it is incumbent on all fire/rescue agencies (and assisting law enforcement and government agencies) to adopt SEMS/ICS and use it every day to ensure that it becomes ingrained in the local, regional, state, and national emergency response cultures.

It should be emphasized that ICS/SEMS is not intended to be placed on a shelf and used only during disaster operations. That is probably the biggest mistake that an agency can make. ICS/SEMS are tools that can (and should) be used to enhance the safety and effectiveness of everyday fireground, EMS, hazmat, and rescue operations. Daily use of ICS/SEMS for smaller incidents will help ensure that all ICs, officers, firefighters, rescuers, and responding agencies are prepared to use these powerful tools when disaster strikes.

Response issues

It's been proven effective in reducing rescuer injuries and fatalities and improving the survivability for trapped victims if fire departments respond with rescue and USAR units on the first alarm whenever there's an indication that their specialized training and equipment may be needed. Generally, it's prudent to dispatch rescue/USAR units on the first alarm to the following types of emergencies to expand and reinforce the capabilities and operational safety of the first responder firefighters:

• *Firefighter Down* (*Mayday*) situations on the fireground, including instances where personnel have become trapped by the collapse of burning structures or high-piled stock, firefighters *through the roof* or *through the floor*, or other fireground entrapments. This also includes firefighters lost or missing during fireground operations, as well as personnel injured in locations (i.e. upper floors, basements, roof tops, etc) where technical rescue methods are required to remove them to a place of safety. The IC should establish at least one RIC whenever personnel are being committed to IDLH environments and other high-risk situations where they may require physical extraction if they become trapped, lost, or injured (Fig. 6–4).

Fig. 6–4 USAR Company Members Assigned as an RIC at a Tunnel Rescue Operation Prepare to Move RIC Equipment to a Strategic Location

• *Rapid intervention* operations to locate and extract personnel who become trapped, lost, or injured during the course of non-fireground emergencies in mountainous terrain, in rivers and flood-control channels, by mud and debris flows or landslides, and in other non-fireground rapid intervention situations.

• Emergencies involving the collapse of buildings, trenches, excavations, tunnels, bridges, and other structures, with causes that include (but are not limited to) earthquakes, explosions, design problems, material defects, slope failure, floods, collisions (i.e. vehicles in structures, train derailment, aircraft crashes), sabotage or terrorist acts, or fire damage.

• Traffic collisions with vehicles over-the-side on mountain roads, bridges, and all other high-angle situations requiring technical rescue methods for insertion of personnel and extraction of patients.

• Victims trapped or stranded on cliffs, hillsides, ice chutes, rocks, waterfalls, trails, and other mountain rescue situations (regardless of whether there are injuries) or any other *person-trapped* situation requiring the use of ropes, cables, helicopters, or other technical means to remove victims to safety and (when necessary) proper medical treatment.

• Aircraft crashes, aircraft fires, and alert standbys for aircraft in trouble during flight.

• Train and subway derailments, collisions, and fires.

• Marine emergencies where people may require rescue from the open water, possibly those who abandon ship to escape fires and capsizings. This includes ship collisions, capsized boats, sinking ships/boats with people aboard, fires aboard occupied ships/boats, watercraft *on the rocks*, aircraft crashes in the ocean, etc. This assumes that the rescue/USAR units are trained and equipped for marine emergencies.

- Traffic collisions and other vehicular mishaps involving reports of people trapped in (or beneath) tractor-trailers, bulldozers, earthmovers, cranes, motor homes, or other large vehicles.

- Victims trapped on bridges, window-washing platforms, cranes, high-rise buildings, and other urban high-angle predicaments.

- Industrial accidents where people are reported trapped in machinery or beneath heavy objects or materials.

- All confined-space rescues, including engulfment, electrocution, entrapment, or injury. Also could involve unconscious or missing victims.

- All reports of people trapped or missing in mines and tunnels, subterranean storm channels, waterworks, and train/subway tunnels. Includes fires, floods, collapses, explosions, and other mishaps within tunnels and mines.

- Victims trapped or swept away in rivers, streams, flood-control channels, flash floods, fast-rise floods, and mud/debris flows.

- Potential suicide and hostage situations in high-risk environments where the safety of law enforcement crisis negotiators requires fire department support. Support may be in the form of belay lines to deal with subjects in high-angle environments or observation equipment such as thermal imaging systems, search cameras, trapped-person locators, fiber-optic systems, etc. Support may involve cutting and breaching to obtain access to subjects and victims. Typical situations include potential *jumpers* or hostage incidents on bridges, buildings, and other high-angle situations, or similar situations in confined-spaces, tunnels, and certain industrial settings.

- Landslides, avalanches, rockslides, and other emergencies caused by geological failure.

- When rescue helicopter personnel require the assistance of additional personnel trained to conduct helicopter hoist operations, short-haul rescues, open-water deployment, helo/high-rise operations, or victim packaging for emergency missions. This assistance could be required in the mountains, on the ocean, on cranes and towers, and for other unusual search and/or rescue missions. This assumes that the rescue/USAR units are trained in helicopter rescue operations.

- Ground-based searches requiring thermal imaging, night vision, trapped-person locators, fiber-optic systems, search cameras, void space search, and other scenarios where victims are missing.

- Helicopter-based searches requiring hand-held thermal imaging systems, hand-held night-vision systems, or other equipment carried on USAR/rescue units.

- High-rise fires. USAR/rescue units in some departments are designated as helo/high-rise teams, capable of being deployed onto the roof-top via helicopter to conduct rooftop ventilation, helicopter evacuation of victims from the roof, SAR on the upper floors using biopack SCBAs, and fire attack.

- Refinery fires where victims may become trapped on cracking towers and other high-angle predicaments and where other USAR capabilities may be used.

- Train fires, for forcible entry, thermal imaging, etc.

- Victims trapped in submerged vehicles.

- Any emergency where the IC determines that USAR task force thermal imaging, shoring, lifting, hauling, searching, cutting, breaching, or other capability may be needed.

It's also the practice of some fire departments to dispatch a rescue company or USAR unit (including USAR task force fire stations) to all multi-alarm structural fires. Some departments assign USAR/rescue units to augment the existing RIC or to conduct structural safety evaluations. The unit may have to be prepared to shore and stabilize structures in danger of collapsing after fire operations, perform specialized tasks such as thermal imaging, or perform other assignments that make use of the specialized capabilities of these units.

ICs should strongly consider continuing the response of USAR/rescue resources until the victim has been extracted and all personnel are out of potential danger zones such as trenches, building collapses, over-the-side areas, etc. As long as rescuers and victims are in the hazard zone, the IC should continue the USAR resources to provide technical advice, act as trained safety officers, provide backup rescue capabilities, and assist with the rescue as needed.

Based on the incident type, the incident text, and other information, the IC should consider the potential need for additional resources. Typical *special resource requests* for USAR incidents include the following:

- *Extra engine or truck companies* (or, in wildland areas, camp crews) to provide extra personnel for manpower-intensive emergencies such as high-angle rescues, trench rescues, structural collapses, confined-space rescues, etc.

- *Shoring or collapse units* for trench rescues, structural collapses, mud and debris flows, heavy extrications involving tractor-trailers, and other incidents requiring extensive shoring and cribbing.

- *Hazmat units* for confined-space rescues, tunnel rescues, and other incidents that require atmospheric monitoring and other hazmat capabilities.

- *Heavy equipment resources* such as bulldozers, loaders, track hoes, cranes, etc., to advise or assist with heavy lifting operations, structure collapse operations, trench/excavation collapse rescues, and other incidents where heavy equipment may be required.

- *Heavy wreckers* to assist with heavy lifting operations, entrapments involving tractor-trailers, stabilization of vehicles in precarious positions, railway accidents, etc.

- *Rescue companies* or *USAR units from other agencies* to assist with large or complex rescue/USAR incidents.

- *USAR canine search teams* for structural collapse, mud and debris flows, avalanches, trench/excavation accidents, water rescues, and other situations involving missing victims.

- *Structural engineers* for technical support during collapse operations.

- *Industrial-sized dirt vacuums, air knives,* and *other mechanical devices* to remove soil and other material from around trapped victims.

- *Crisis negotiation teams* to deal with subjects threatening to jump from bridges, buildings, towers, or other high-angle locations. Teams to deal with hostage situations and to address other emergencies that fall under the purview of crisis negotiators.

As necessary, consult via radio, MDT, or cell phone with USAR resources for recommendations on other special resources. Consider the response time of USAR/rescue resources and other special units to your incident. In some cases, it will help to expedite their response using helicopters. Consider the need for helicopter hoist operations and other aerial capabilities and request them early in the incident to reduce *reflex time.*

At the scene

The following section describes actions that should be taken by the USAR/rescue IC at the scene of an incident (Fig. 6–5). Conduct an eight-sided size-up to assess (and then report) the following conditions:

Fig. 6–6 Firefighters Searching for Trapped Victims in an Occupied Parking Structure that Collapsed during the Northridge Earthquake

Fig. 6–5 Rescue Officers Sizing up the Scene of a Train that Derailed in a Residential Area of L.A. County

The exact location of the incident and the best access for ground units and helicopters (Fig. 6–6).

At a scene like this, an eight-sided size up is required to ensure that you have observed critical factors and cues that will help you set the stage for effective operations.

- The nature of the victim's predicament.

- Is the victim injured?

- If the victim falls, is dislodged from his present position, or if debris rolls/falls onto him, can he be injured or his entrapment worsened?

- Is the victim able to assist in his/her rescue?

If trapped on a cliff, in a trench, or a similar predicament that is at first inaccessible to rescuers, can the victim move (under his own power) to a safer location/position until you establish a rescue system and get a rescuer to his side? (See Fig. 6–7.) What factors should be evaluated during the size up of this emergency? What dangers need to be addressed? When does the size up end?

Fig. 6–7 Firefighters and USAR Company Members Stabilizing the Walls of an Excavation that Collapsed and Buried a Construction Worker

Fig. 6–8 Before Attempting a Rescue like This, it's Important to Conduct a Rapid Size-up

- Is this situation best handled with a helicopter hoist rescue, a ground-based rescue, or some combination thereof? (See Fig. 6–8.) A rapid size-up determines a number of factors like whether a ground-based operation or a helicopter-based rescue is most appropriate for this particular situation.

- Are any victims missing?

- What is the exact extrication problem?

- If a vehicle is involved, does it require stabilization (i.e. cables, ropes, etc.) to prevent movement?

- If a damaged structure or trench is involved, what stabilization measures are required to prevent secondary collapse?

- Begin stabilizing the incident and/or victim as much as possible until he or she can be extracted.

- Ensure, whenever possible, continued medical treatment of trapped victims throughout size-up, stabilization, and extrication operations.

- Consider the need for a helicopter to stand by for immediate aerial transportation of the victim following extrication.

- Recognize critical cues and life hazards for victims as well as rescuers, including those that indicate the need for secondary responders (USAR task forces) to help manage them.

- Initiate effective first responder operations within the scope of their training and equipment.

- Recognize the boundary between the safe limits of first responder and secondary responder operations and understand when it's best to concentrate the efforts of first responders on stabilizing the situation as best they can until the arrival of secondary responders.

- Assume command and name the incident.

- Begin stabilizing the incident and/or victim as much as possible until he or she can be extracted.

- Request additional resources based on the need for specialized equipment and trained personnel to conduct stabilization, search, rescue, and rapid intervention operations.

- Develop a strategy to perform the rescue if on-scene personnel have the proper training and equipment to conduct the operation safely and effectively.

- If the incident exceeds the capabilities of first responders, develop a strategy to protect the patient and rescuers then stabilize the incident until the arrival of USAR/rescue units. If necessary, isolate the danger zone and deny entry until USAR units arrive. Consider the need to provide indirect assistance to the patient (i.e. lowering an oxygen mask and helmet to the patient, providing ventilation, preventing secondary collapse, etc.).

- Consider the need for unified command on large or complex incidents.

- Be prepared for long-term operations. Because of unforeseen complications, many technical rescues take twice as long to safely complete than initially anticipated.

- Consider the need for a helicopter to stand by for immediate aerial transportation of the victim following extrication.

- Consider rotating crews and providing rehab to reduce fatigue and maintain good working strength until the rescue is accomplished.

Implement Rescue Plan/Incident Action Plan (IAP)

After conducting the eight-sided size-up and report, it is time to implement the IAP as follows:

1. Develop a strategy to perform the rescue if on-scene (first responder) personnel have the proper training and equipment to conduct the operation safely and effectively.

2. If the complexity and technical demands of the incident exceed the capabilities of first responders, develop a strategy to protect the patient, protect rescuers, and stabilize the incident until the arrival of USAR task forces. If necessary, isolate the *danger zone* or the *IDLH zone* and deny entry until USAR units arrive. Consider the need to provide indirect assistance to the patient (i.e. lowering an oxygen mask and helmet to the patient, providing ventilation, preventing secondary collapse, etc.).

3. If the response of USAR/rescue units is extended, consult with rescue officers via radio or MDT regarding special needs, including the possibility of expediting the response of rescue specialists via helicopter.

4. Request additional resources early to reduce the *reflex time* (a measure of the time between calling for a resource and the arrival of the resource) and thereby avoid critical delays.

5. Consider the use of unified command on large or complex multi-agency incidents.

6. Be prepared for long-term operations, as unforeseen complications may necessitate that technical rescues take two or three times longer to complete than initially anticipated.

7. For manpower-intensive operations, consider rotating rescuers every 30 minutes (or another appropriate time frame) and providing rehab to reduce fatigue and maintain good working strength until the rescue is accomplished.

8. Apply lookout, communication, escape route, and safety (LCES) principles to the operation.

Applying LCES to rescue operations

The concept of LCES was originally developed by wildland firefighters as a guideline to avoid common mistakes that result in fireground fatalities. For years LCES has been a mainstay of wildland firefighters whose every move is influenced by it. LCES figures into the development of strategy and the employment of tactics.

Progressive fire departments have adapted LCES for use in non-wildland situations such as interior structure firefighting, terrorism incidents, hazmat response, and rescue. LCES as a rescue concept became prominent when members of FEMA USAR task forces included it in their operational action plans at the Oklahoma City bombing in 1995. In places like L.A. County, LCES is commonly used in all forms of high-risk rescue emergencies (Fig. 6–9).

Fig. 6–9 An IC Issuing Instructions at the Scene of a Major Collapse SAR Operation

Large-scale rescue incidents may require the response of multiple agencies like law enforcement, public works, heavy equipment, state and federal SAR resources, and others. It's important for the IC to be prepared to orchestrate the operations of these disparate agencies and resources, to maintain control of the scene, and to ensure that operations are conducted with a reasonable margin of safety and personnel accountability.

Before committing personnel to the danger zone of a rescue scene, the IC should always ensure that LCES has somehow been addressed and that all members are aware of them. The following is a brief review of LCES as it applies to rescue.

Lookout. Some member of the team (or a reliable responder such as a firefighter, police officer, structural engineer, construction worker, public works member, etc.) should be assigned to observe the rescue scene. They should be instructed to look for signs of impending secondary collapse, secondary explosion, fire, frayed ropes, avalanches, rockslides, flash floods, mud and debris flows, or other immediate life hazards that could kill rescuers. It may be necessary to place the lookout in the basket of an aerial platform, on an aerial ladder, on an adjacent building, on a mountainside, or even in a helicopter to ensure that they can view the entire rescue scene (Figs. 6–10 and 6–11). This melding of different resources (e.g. rescue and hazmat) to perform complex operations is typical of an effective response to some rescue emergencies. As part of the IC's insistence on the use of LCES at the scene of the Pentagon collapse following the 9-11 terrorist attacks, a structural engineer and a search team member from a FEMA USAR task force were assigned to a personnel basket around the clock, observing for signs of secondary collapse and for trapped victims who might be uncovered during *selective debris removal* operations.

Fig. 6–10 Rescue and Hazardous Materials Officers in Supervisory Positions during Confined Space Rescue Operations

Fig. 6–11 Use of LCES at the Pentagon Collapse following the 9-11 Attacks

It may be necessary to post more than one lookout. The incident may also require the use of theodolites, plumb bobs, and other tools that can indicate movement of a building preceding a secondary collapse. PS-wave detectors can be mounted to identify earthquake aftershocks before their destructive waves arrive and cause secondary collapse, thereby giving rescuers a few seconds to get into safe refuge areas or safety zones.

Communications. Every rescue operation should have a clear communication plan that includes designated radio channels for certain functions/teams. However, the communication plan must extend beyond the use of radios, which are subject to failure and which can be lost or damaged during the course of rescue operations. Firefighters engaged in rescue must be familiar with other forms of communication such as whistles, air horns, hand signals, and—of course—voice commands. All personnel operating in and around the rescue site should be familiar with the communication plan, and each officer should ensure that his firefighters are using the components of the plan appropriately (Fig. 6–12). What LCES-related concerns and needs are evident here?

Fig. 6–12 L.A. County Firefighters and USAR Units Searching for Trapped Victims at an Incident that was Spread over Several Blocks in a Residential Neighborhood

Clear position designators are also critical to communications. The use of identification vests, properly marked helmets, armbands, or other identifiers should be mandatory. In a disaster setting when these methods may not be available to everyone (including personnel who report to duty from home), the use of marker pens to hand-print designations on shirts, helmets, or even on arms, is preferable to the chaos that occurs when everyone looks the same and no one can identify who's responsible for what. In disaster-prone areas, pre-designated caches of armbands, helmets, and vests can assist in the process of communication. Hard-wired communication is sometimes needed in environments such as tunnels, mines, confined spaces, and collapse zones where radio waves will not penetrate (Fig. 6–13).

Fig. 6–13 ICS Position Vests Help to Identify Who is in Charge of What Assignment and Reduces Confusion during the Most Critical Operations

Communication also includes the use of clear and concise IAPs that coincide with what's actually happening in the rescue zone. In other words, it's helpful if the IAP matches (to a reasonable degree) the actual conditions on the scene. Too many times firefighters will look to the IAP for guidance and discover that it's either outdated or inaccurate or does not begin to convey the actual operations that are occurring. IAPs can be a great help if they are accurate and well thought out. Conversely, poorly developed IAPs that don't match the reality on the ground can sometimes make the situation worse by misdirecting tactics and strategy.

Escape route. Every firefighter engaged in high-risk rescue and every supervisor and the IC if possible, should have a clear idea of the primary and alternate escape routes. Every officer should brief his crew or team on the chosen escape routes during each entry into an IDLH environment. The escape routes should be reevaluated as conditions change. If necessary they should be revised (Fig. 6–14).

Fig. 6–14 In Some Technical Rescue Situations, Rapid Escape May Not be Feasible in the Normal Sense

- What is the escape route here?

- How will personnel know that it's time to make an emergency escape?

- How does the IC ensure that everyone is clear about the signal to start an "operational retreat" through the appropriate escape route(s)?

Escape routes should be the fastest, safest way out of the danger zone or the fastest and most direct way to a safe refuge. In the event of a secondary collapse, fire, explosion, flooding, or other unexpected secondary event, preplanned escape routes may save the lives of firefighters and other rescuers. Escape route(s) should be identified by fluorescent spray paint markings or with signs, fireline tape, lumber crayons, and/or other clearly identifiable method (Fig. 6–15).

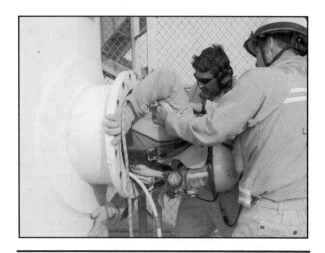

Fig. 6–15 Is This a Good Escape Route?

In confined space rescue operations, there may not be any other choice for escape (which makes it even more important to look after the other components of LCES). Take precautions that prevent the need for an emergency escape.

Escape routes can take different and sometimes unique forms. During SAR operations following an earthquake that shook the Philippines in 1992, USAR specialists from Dade County (Florida) and Fairfax County (Virginia)[1] found that rapid escape through the corridors of an overturned hotel was infeasible during aftershocks. To expedite egress from the collapsed building, rescuers stacked mattresses outside windows. The agreed-upon escape plan required rescuers to simply scramble to the designated windows and dive out one at a time, each rescuer rolling off the mattresses just in time for the next team member to land safely (Fig. 6–16).

Fig. 6–16 Might This "Confined Space Crawler" Represent a Component of an Effective Emergency Escape Route?

Stacking mattresses as an escape route might seem comical to some who've never operated inside a collapsed building with aftershocks continuing to strike, but it was clearly a simple and workable plan. In fact, the plan was successfully used to evacuate rescuers from the collapse zone numerous times over a period of several days. When faced with unusual conditions, it's important for team leaders and officers to *think outside the box* when addressing the safety needs of their fellow rescuers (Fig. 6–17).

Fig. 6–17 A Potential Escape Route, Viewed from within a Collapse Zone in the Pentagon after the 9-11 Terrorist Attacks

Safe zone. Team leaders and officers should identify at least one safe zone, an area safe from secondary collapse and other hazards, into which rescuers can retreat in the event of an aftershock, explosion, secondary collapse, or other unplanned event. The safe zone may be outside a building beyond the collapse zone, usually the same distance as the height of the building, or it may be beneath a rock ledge, inside a building, or on a hillside. Fig. 6–18 shows an example of a safety zone created by shoring an unstable section of a vertical shaft where L.A. city firefighters are digging to rescue a construction worker who was buried when the shaft collapsed.

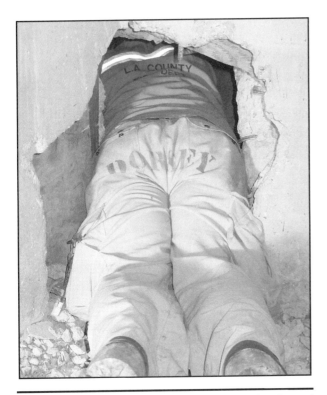

Fig. 6–18 Sometimes the Escape Route must be Created such as Breaching and Tunneling in a Collapse Zone

If the rescue operations are occurring inside a building, trench, excavation tunnel, or other structure, and escape to the outside will take too much time or is otherwise infeasible, the safe zone may be designated within a stairwell, air shaft, shored area, or other fortified space.

In some cases, it may be necessary to construct a safe zone inside a damaged structure, fortifying it through the use of shoring, cribbing, or other method. Everyone entering the danger zone should be clearly aware of the safe zone(s). In the case of an unplanned event, the team leaders or officers should conduct head counts at the safe zones to ensure that all rescuers made it to safety and to determine whether or not anyone needs assistance.

Considerations of the Safety Officer

A primary concern of the IC during emergency incidents is the need to provide a reasonable margin of safety for firefighters and other public safety personnel. If the IC cannot ensure a reasonable degree of safety for those who are actually doing the work, successful outcome of the incident will be jeopardized. The designated safety officer plays a critical role in this process, but his effectiveness will be highly dependent upon the officer's ability to quickly recognize operational hazards and recommend corrective action in time to prevent unnecessary injuries to personnel. This is directly related to the safety officer's knowledge and experience of the hazards common to the type of operation being performed. Nowhere is this more evident than when a department is confronted with an unusual technical rescue incident (Fig. 6–19).

Fig. 6–19 What are Your Concerns?

Assume you are the Safety Officer at this incident involving a worker buried nearly 40 feet beneath a deluge of sandy soil that collapsed into this shaft during a rail construction project. What are your concerns as you observe this firefighter working to remove soil with the assistance of a hydro-vac?

Even without formal safety officer certification, the typical fire department officer should be capable of assuming the duties of the designated safety officer during typical fireground operations[2] because of his experience with fire behavior and his understanding and experience with the tactics and strategy being used to control the fire. In these cases, the safety officer's years of personal experience with fire conditions and firefighting tactics combined with common sense and the ability to adapt to changing conditions are his greatest allies. With experience, the officer intuitively recognizes situations that shout *watch out*, and is able to objectively weigh the risk vs. gain equation that is driving the decisions of the IC, the division/group supervisors, and individual officers at the incident.

Conversely, this is where the safety officer can get into trouble when confronted with an unusual incident with which the officer has little or no experience. Without personal experience to guide them, safety officers may be somewhat handicapped. They may not immediately recognize situations that are IDLH cases. They may not anticipate changing conditions that are typical of the incident type. In addition, they may not be aware of laws that mandate certain types of equipment and procedures to ensure worker safety during management of the incident. In some cases, there may even be criminal penalties for failure to observe certain worker safety provisions. Worse, failure to recognize and mitigate the hazards may result in preventable injuries or death to firefighters and other rescuers, or the victim (Fig. 6–20).

Fig. 6–20 What are Your Concerns?

As the Safety Officer of this incident involving a worker buried in an excavation, what are your chief concerns? How will you address them?

Urban SAR incidents are notorious for their unusual nature, or at least for the infrequency that makes them appear unusual. Depending on where you work, USAR incidents such as trench collapse or swift-water rescue may be *once-in-a-career* events.[3] When facing such situations, the formal training of safety officers and their personal planning and preparation will take on primary importance.

Worker safety laws and regulations related to rescue command

It's not practical to enumerate and define every law that's applicable to urban search and technical rescue. To do so here would be somewhat impractical because laws and regulations change more frequently than books are published, and consequently, some of the laws and regulations listed here would be antiquated within a few years. Nevertheless, it is important for ICs and rescue officers to recognize certain laws and regulations—recognizing that they may change before the reader has this book in their hands—to infer new trends in worker safety and public safety laws that will affect rescue operations.

Recent firefighter fatality incidents resulting in investigations and fines against fire departments for failure to comply with applicable standards, laws, and codes demonstrate that fire officers and agencies are being held to account for certain serious violations.

For example, when three Oregon firefighters died in a commercial blaze in November 2002, the Oregon division of OSHA launched an investigation that led to the involved fire department being cited for 16 violations of federal safety standards (13 classified as serious) and fined more than $50,000.[4] The fire chief, who had previously been praised for ordering an operational retreat after noticing a spongy roof on the burning auto supply store, was the subject of serious criticism by OSHA investigators. The chief was criticized for a broad range of violations, including the following:

- The ICS used by the local fire department did not meet the requirements of NFPA standards; there were breakdowns in communications and no RIC was established (fined $7,000).

- At least two firefighters (standby members) were not assigned outside the building in a state of readiness to assist firefighters operating inside when they became trapped (fined $5,000).

- A personnel accountability system was not in evidence (fined $5,000).

- Firefighters did not receive annual fit tests for SCBA face pieces (fined $13,000).

- SCBA was not used by one firefighter assigned to roof operations (fined $2,500).

- The city did not ensure that each firefighter could effectively use a SCBA during an emergency (fined $5,000).

- SCBA were not inspected daily (fined $500).

- Other violations of varying severity for a total fine of $50,450.

As of this writing, the local fire department is considering an appeal of the citations, and criminal and civil court proceedings are pending. The fire chief who commanded the incident is in legal jeopardy for the cited infractions and other repercussions. It's increasingly clear that when firefighters die in the line of duty, supervisors may be held accountable if they do not comply with appropriate rules, laws, and standards. Some key regulations and standards related to rescue include the following:

- NFPA 1670, *Standard on Technical Rescue*

- NFPA 1006, *Standard on Professional Competencies for Technical Rescue*

- NFPA 1983, *Standard on Life Safety Rope, Harnesses, and Hardware*

- NFPA 1951, *Standard on Protective Ensemble for USAR Operations*

- NFPA 1470, *Standard on Search and Rescue Training for Structural Collapse*

- NFPA 1500, *Standard of Fire Department Occupational Safety and Health Programs*

- NFPA 1561, *Standard on Fire Department Incident Management Systems*

- OSHA 1910.134, *Respiratory Protection*

- OSHA 1926.650, *Trench and Excavations*

- OSHA 1910.146, *Permit Required Confined Spaces for General Industry*

- Standard operating guidelines (SOGs)

It's been said that effective training and realistic SOGs provide the foundation for effective management of routine emergencies. Conversely, it is also generally recognized that unusual incidents often exceed the scope of written guidelines, challenging emergency responders to rely more heavily upon their experience and critical judgment. In both instances, a subtle but important part of this equation is the ability of firefighters and rescuers to make the most effective use of *available resources* sometimes considered unorthodox. In other words, they often need to *improvise*. Those who lack the ability to improvise during the course of unusual emergencies will be hampered in their efforts to save lives and property. History is replete with examples of unusual emergencies in which strict reliance on rules and regulations simply didn't get the job done, compelling firefighters and rescuers to improvise the use of available tools, equipment, and personnel in unusual ways.

Periodically an incident comes along that compels public safety personnel to deviate from their normal repertoire of tactics, strategies, and equipment. If they find themselves unable to adapt to fast-changing conditions, public safety personnel can find themselves in lethal danger—and quickly. Adaptability is part of the culture of the fire and rescue services, and therefore, unusual emergencies represent not just serious challenges to overcome, but also ripe opportunities to devise and master new strategies, equipment, and methods (read P-R-O-G-R-E-S-S).

Rescuer safety

The safety of firefighters and other rescuers is a key factor in the success of a rescue operation. Not only is firefighter/rescuer entrapment, injury, or death during rescue operations a tragedy in itself, but it also causes a cascading effect that reduces the chance of survival for the original victim(s). On the fireground, many fire departments rely on rescue companies and USAR units to perform rapid intervention operations or to augment existing rapid intervention capabilities (Fig. 6–21).

Fig. 6–21 Final Safety Check

A final safety check should be required before personnel enter an immediately dangerous to life and health (IDLH) environment. And rightly so. The FDNY and many other fire/rescue agencies across the United States have proven that fire companies specializing in rescue are one of the best assurances that firefighters will be located, accessed, and extracted in a timely manner if they become lost, trapped, or seriously injured on the fireground.

A growing number of fire departments across the United States have made it standard policy to dispatch firefighter-staffed USAR companies or rescue companies to all multiple-alarm fires. They do so for the sole purpose of upgrading first-alarm rapid intervention resources by providing units with specialized search and extrication equipment operated by personnel specially trained and highly experienced in rescuing people (including rescuing other firefighters in fireground situations). This policy of deploying specially trained/equipped fire department rescue/USAR units to rescue firefighters who become lost, trapped, or injured while battling fires represents a revolution in rescue (Fig. 6–22).

Fig. 6–22 One of the Biggest Challenges in Rescue is That of Protecting Rescuers Engaged in High-risk Operations

Every firefighter and rescuer should be considered an unofficial *safety officer*. They should be constantly alert to danger, always playing the *what if* game and prepared to alert other members to take protective action in the event of an unanticipated secondary collapse, secondary explosion, earthquake aftershock, or fire, etc. Even when there's a designated safety officer at the site of a rescue, every other firefighter/rescuer should consider himself a safety officer in his own right. If we have this mindset, we are more vigilant for situations that can lead to the entrapment, injury, or death of our colleagues and ourselves.

Although on average more than 100 firefighters are killed in the line of duty every year in the United States and significant numbers die in other nations annually, the fire service does not operate with a doctrine of *acceptable losses*. The goal among most fire service officials is to continue reduction of firefighter and rescuer fatalities as much as possible, while

still getting the job done. As firefighters and rescuers, each of us accepts the risk of being killed in the line of duty, and we understand that unforeseeable circumstances sometimes combine to place us in untenable situations where life or death is a matter of seconds and degrees. We also realize that seemingly inconsequential events can have catastrophic results.

We also understand that in some cases the only way to rescue a trapped victim is to place the lives of rescuers on the line. We are compelled by our jobs to consider committing personnel into dangerous situations to conduct SAR operations during daily emergencies and disasters alike, where—if something goes wrong—the result may be tragedy. Each of us must be prepared to make that most critical individual decision when confronted by situations of that nature.

This is not to say that rescue should be considered a suicide mission. To the contrary, when rescuers are lost, seriously injured, or killed, it sometimes means that something went wrong and that it might have been preventable. Every firefighter and rescuer should be mentally and physically prepared to rescue other members who need lifesaving assistance. Likewise, every rescue plan should take into account the potential for rescuers to get into trouble and should subsequently include a rapid intervention plan (Fig. 6–23).

Fig. 6–23 LACoFD USAR Company at the Scene of a Three-alarm Commercial Building Fire

The department has adopted the practice of assigning USAR units to augment the RIC on all multi-alarm fires. Because of their experience dealing with structure collapses and other technical issues, USAR firefighters at greater alarm fires are also asked to assess structural stability and to assist the safety officer in observing for impending trouble. In recent years, this approach has resulted in the rescue of at least one firefighter who might otherwise have been lost in fireground entrapment situations.

Rescue operations inherently involve some level of calculated risk. The key is to reduce the risks to a reasonable level whenever possible through effective planning, training, equipment procurement, and rational emergency response. One of the advantages of establishing a formal rescue program is that a fire department (or group of fire departments in the case of regional systems) can reduce the level of risk. A rescue program ensures that rescue units are able to conduct rescue operations effectively with a reasonable degree of safety. Preparedness should be based on solid planning, training, proper equipment, research and development, and implementation of time-proven tactics and the employment of good strategies.

Formal rescue systems can serve as clearinghouses for valuable rescue information. They can be sources of unique research and development to improve equipment, training, and operations, and they can become repositories of years of experience among rescue unit members. Formal rescue systems can provide technical support for ICs who may not be familiar with the details and needs of particular types of rescue emergencies and operations, but who nevertheless understand the effective use of specialized resources to get unusual jobs done. This is particularly true when it comes to firefighter and rescuer safety.

Operational retreat

All firefighters and rescuers, and especially ICs and officers, must be familiar with the signal to begin an *operational retreat*. An operational retreat involves the immediate exit of the danger zone or building. Operational retreats are initiated by a standard signal given when secondary collapse or some other secondary event is imminent or when another immediate life hazard has been discovered. An operational retreat system is necessary for the IC to ensure that all personnel are safe and accounted for and to determine whether any rescuers are missing, trapped, or injured (Figs. 6–24).

Fig. 6–24 USAR Firefighters Assigned as an RIC during Swiftwater Rescue Operations

In this instance, the focus is on "downstream safety." The concept of rapid intervention has always been used in swiftwater rescue, confined space rescue, and other technical rescue disciplines.

An operational retreat at suspected terrorist attacks should not be conducted in a way that allows terrorists to predict the location of rescuers after they exit the operations area after multiple bomb threats or other hazards. In Figure 6–25, rescuers repeatedly assembled in the same place following several operational retreats at the Pentagon collapse prompted by multiple phone threats. What might terrorists be looking for if they employ this technique?

Fig. 6–25 Is it Appropriate to Use Such Holes as Escape Routes?

Fig. 6–26 Operational Retreat at the Pentagon Collapse Operation after a Terrorist Bomb Threat was Received

In this figure, Firefighters and rescuers worked the 9-11 Pentagon attack, where the airliner "punched out" after traveling through the outer ring of the building. Is it appropriate to use such holes as escape routes? Are additional safety precautions needed?

Rapid intervention

Nowhere else are the rescue skills of firefighters more important then when other firefighters and rescuers become lost, trapped, or injured at the scene of a rescue or on the fireground. Rapid intervention is the most basic and essential form of rescue. It is intended to save the lives of those who have voluntarily placed themselves in harm's way to help others. Rapid intervention is important also because we know that effective and timely rescue of downed or lost firefighters enables other fire/rescue personnel to resume their original mission of saving the lives of citizens and protecting property and the environment.

In Figure 6–26, personnel accountability was assessed after an "operational retreat" was ordered because of telephoned bomb threats at the Pentagon following the 9-11 attacks.

In Figure 6–27, An L.A. County FD captain fell through the hole (in background) in the floor of this commercial building and landed in the basement during a two-alarm fire the night before this photo was taken. He nearly died from exposure to combustible products and heat before he was rescued by an RIC consisting of a USAR task force and one engine company, augmented by a second engine and a truck company.

The captain was unconscious, his SCBA and turn-out jacket ripped off during the fall between floors by nails protruding from the joists. He was burning and inhaling fire-heated gases. Firefighters had to breach a wall to reach him while others stood their ground with hose lines to protect the rescuers as the well-involved commercial building threatened to collapse on all of them.

Fig. 6–27 This Hole Burned through the Floor Lead Straight to the Basement for an Unfortunate Captain

In Figure 6–28, moments after the successful rapid intervention operation, the injured and burned captain (white T-shirt) is being treated by firefighter/paramedics who were assigned as medical group once the "firefighter down/rapid intervention" report was announced. Seconds after this photo was taken, the roof collapsed, burying the area where the captain had been trapped.

Fig. 6–28 Fire Captain being Treated after Rescue by a USAR Company Acting as an RIC

The manpower required to rescue this single firefighter (four companies) is not unusual. Recent testing conducted by the Phoenix Fire Department has demonstrated that it typically takes multiple companies to rescue one downed firefighter, especially if his location is not known or if he is buried by collapse debris. Assigning USAR and rescue units to rapid intervention duty is one way to take advantage of their special training and capabilities.

In Figure 6–29, the hole at the top is where the captain fell through the floor. The hole through which this firefighter is looking was opened by the RIC using chain saws and hand tools. This allowed two firefighters to crawl into the basement and remove the downed captain through the same hole. From there, the semi-conscious firefighter was carried through the burning commercial building to safety, just moments before the roof collapsed.

Fig. 6-29 Holes Cut for Access

When a rapid intervention situation occurs, the priorities of emergency operations must immediately change, and the IC must quickly identify and communicate the new priorities. The IC must identify what has just happened and what secondary events are likely to occur unless stabilization precautions are taken. The IC then must take effective actions to ensure that one or more RICs are immediately engaged in operations to locate and extract any trapped firefighters and rescuers.

Firefighters may be forced to halt their original assigned tasks to assist trapped, injured, or lost colleagues. In some cases this has the unfortunate effect of diverting help away from the original victim(s), at least temporarily. Aside from the potential tragic consequences to the trapped/lost firefighter or rescuer, the effect on the original victim is yet another reason to avoid the need for rapid intervention operations by taking prudent precautions.

Unless the IC has wisely ensured the immediate availability of an RIC with a solid RIC plan and additional resources at his disposal in case of an adverse event, the situation can quickly spin out of control. Furthermore, entrapment or death of rescuers may cause hesitation on the part of others who may (wisely in some cases) choose not to place themselves at risk of the same fate.

A rapid intervention situation can reduce the overall effectiveness of rescue operations by reducing the effectiveness of rescuers in trouble and also by causing additional mental trauma to their team-mates. All these factors add up to a reduced chance of survival for the original victim(s). The best solution is to prevent rapid intervention situations whenever possible through the use of effective strategies and tactics that recognize accepted *risk vs. gain* equations and use reasonable precautions to ensure the safety of firefighters and other rescuers.

If you as a rescuer become lost, trapped, or injured under these conditions, you should expect that the IC will immediately mount a rapid intervention operation to locate and rescue you from harm before normal SAR operations resume, or simultaneous to other SAR efforts. If you aren't sure that the IC has made provisions for this, you may want to verify that these measures are in place before your crew is committed so far into the IDLH zone that escape may be impossible if something goes wrong. In Figure 6–30, recruit firefighters practice rapid intervention techniques during a firefighter safety and survival drill at a training academy.

Fig. 6–30 A "Firefighter Safety and Survival" Drill at a Fire Academy

Conversely, all firefighters and rescuers should be prepared to conduct rapid intervention operations to rescue other team members who become lost, trapped, or injured during the course of collapse SAR operations. Rescue companies and fire department-based USAR units should be prepared to conduct timely rapid intervention operations on the fireground or to augment existing RIC resources. All ICs should be prepared to manage RIC operations. It should be standard protocol for the IC to designate an RIC, develop a rapid intervention plan, and ensure sufficient resources are in place before committing crews to an IDLH situation.

Whether at the scene of a technical rescue or on the fireground, the designated RIC should be in a full state of readiness outside the IDLH zone, dressed in full *battle gear* to immediately launch a rapid intervention effort if something goes wrong. In a disaster setting, we all understand that the IC or the officer in charge (OIC) may not have the luxury of designating RIC teams whose only role is RIC. All the same, the IC or OIC should still have a plan to rescue personnel who become lost, injured, or trapped, even in the middle of a disaster.

Generally, an RIC plan should include a radio channel for RIC operations, an RIC team officer, protocol for conducting SAR in the particular environment at hand, and provisions for augmenting the RIC with additional personnel and equipment as necessary. There should be a medical group ready to take over treatment of the RIC subject once they are removed from the IDLH zone. The medical group should include paramedics and an ambulance and perhaps even a medevac helicopter on standby for high-risk operations.

There should be a standard radio designation for RIC operations as well as for other notification and communication protocols. When a rescuer is found in need of assistance, a standard radio call such as *rescuer down*, *firefighter down*, or *mayday* (depending on the individual agency's protocols) should be issued. The IC or OIC should call for radio silence to ascertain the downed rescuer's location and predicament, and an RIC operation should be launched immediately using the designated RIC channel.

In short, rapid intervention is not just for confined-space rescue and interior firefighting. There should be an RIC capability for every type of rescue in which firefighters and other rescuers are in danger of becoming lost, trapped, or badly injured (requiring physical extrication) if something goes wrong.

Basic Rules of Commanding Rescue Operations

Physicians have a rule about the ethical treatment of patients that essentially goes like this: Do no harm to the patient during the process of evaluating and treating him or her. Likewise, firefighters and other rescuers should have an oath that defines our responsibilities when responding to emergencies and disasters involving lost or trapped victims. It should begin with the refrain similar to that of the physician: *Do no harm to the victim during the process of locating and rescuing him or her, and conduct the rescue in such a way that rescuers are not unnecessarily lost.*

Upholding such an oath requires prudence, common sense, good training, a rational mind, and the ability to figure things out. It sometimes requires ICs and officers to detour from the normal approach. Above all, it requires an overriding concern for the patient with a reasonable concern for the safety of the rescuers. It requires that we place the welfare of the victim above pride and our sense of self-sufficiency. Here are some other basic rules of rescue:

Rescue rule 1: have a preplan and use it

Effective rescue commanders develop rescue preplans for local hazards long before the alarm sounds. It may not be feasible to create a specific rescue plan for every trench being dug, every construction project getting underway, or any other specific (sometimes temporary and often changing) rescue hazard in the making. However, effective rescue commanders make it their business to plan their general approach to the types of rescue emergencies that are likely to occur. The use of preplans and standard operating guidelines can be helpful in organizing a search and rescue operations at large or complex incidents like this.

As an example, in a typical city at any given time, there may be public works projects involving trenching to lay sewer line, repair piping, etc., and there may be a multitude of construction projects where trenching will be conducted. Knowing this, it's critical that local rescue commanders develop a plan in their own mind (and/or on paper or computer) to address rescue emergencies resulting from trench collapses. The problem with many preplans is that they are often forgotten in the heat of an emergency. Therefore, it's also important to stress the use of preplans as *guidelines* for rescue operations.

Rescue rule 2: call for additional help at the first sign that it's needed

One common mistake made by first responders is to refrain from requesting additional units and assistance from specialized units until they reach a dead end and have no other option. In some cases this has delayed the rescue of innocent victims who put their trust in the first responders to do the right thing and to act in their best interest. In other cases the delay has made the job of the first responders that much more difficult and complicated. In still other cases the delay has placed firefighters and other rescuers in unnecessarily excessive danger.

Figure 6–31 shows USAR company personnel approaching the portal of an abandoned gold mine in the early stages of an operation to rescue a man trapped after falling onto a ledge in a 900-ft vertical shaft within the mine. Note the number of companies (at least 11, including 1 USAR/shoring trailer) parked on the access road in the background. This successful rescue, complicated by conditions within the mine, eventually required seven hours and dozens of personnel. The IC recognized the need for specialized resources and support units early in the operation and requested them immediately, reducing the *reflex time* and facilitating the most timely rescue. The need for this level of resources at difficult technical rescues is not unusual, and the IC should plan for it accordingly.

Fig. 6–31 Rescue in a Gold Mine

The late deputy chief of the FDNY, Ray Downey, was a strong proponent of reducing reflex time. To reduce reflex time, the IC should request additional units at the first sign that they are needed.

Rescue rule 3: don't cancel expert help until the rescue is completed and firefighters/rescue personnel have exited the high-danger zone

Another common mistake among first responders is to cancel incoming truck companies, rescue/USAR companies, and other specialized units before the victim's rescue is assured and before rescuers have exited the danger zone. There are countless examples of ICs who have wrongly assumed that they could complete a difficult rescue without assistance from other units. As a result, ICs have canceled the responses of rescue companies and USAR units only to be forced to re-request those very same units to re-respond when it became apparent that the victim was not going to be rescued without the additional help.

Figure 6–32 shows apparatus lined up at a collapse incident involving numerous structures spread over several residential blocks. Within the first moments of the emergency, the IC created a USAR Branch and assigned a USAR-trained chief officer to coordinate the overall search and rescue operations in three geographic divisions, with USAR unit officers supervising the SAR tasks under the respective division supervisors. This approach made the most effective use of the technical knowledge, experience, and skills of rescue resources in a timely manner.

Fig. 6–32 Apparatus Staging at Collapse Incident

There are also instances where ICs have canceled rescue-trained, back-up units before the rescue was completed only to have a firefighter or some other rescuer become trapped or badly injured when things went wrong. By the time the ICs requested that back-up units re-respond to the mishap, precious time had been lost. In at least one case known to this author, canceling the assisting units led both indirectly and directly to the death of a firefighter when a series of mishaps occurred for which the IC had not planned.

In Figure 6–33, L.A. County F.D. USAR firefighters and a mine rescue team lower a rescuer into a 900-ft vertical shaft inside an abandoned gold mine to extract a trapped victim. Figure 6–34 shows the view looking down the vertical shaft into which the victim fell. The red and white striped rope is the "main line" attached to the rescuer, who has been lowered into the shaft. The black striped rope (around the wood beam) is the rope used by the victim. This is an indication of just how dangerous technical rescues can be and how thin can be the margin for error. This emphasizes the need for the IC to rely on subject matter experts (rescue and USAR personnel) when making strategic and tactical decisions. In Figure 6–35, a backup rescue/litter team consisting of firefighters from an L.A. County F.D. Hazardous Materials task force stand by to carry the victim from the mine after he is extracted from the vertical shaft. L.A. County FD hazmat personnel are trained in confined space rescue operations and are assigned to the First Alarm for confined space and mine rescues, as well as structure collapses and other situations where personnel must work in confined areas.

Fig. 6–34 Vertical Shaft after Rescuer has been Lowered in to Conduct a "Pick-off" Rescue (Courtesy Richard Meline)

Fig. 6–35 Backup Rescue Team (Courtesy Richard Meline)

Fig. 6–33 Lowering a Rescuer (Courtesy Richard Meline)

Anecdotal evidence suggests that canceling assisting units while personnel are still engaged in high-risk operations in IDLH conditions has caused more than one rescuer death in recent years. We certainly should agree that it's not prudent for ICs to cancel backup units until all life-threatening hazards have been addressed, the victim's rescue is assured, and the IC has positive confirmation that resources at the scene are capable of managing any potential mishap. This level of redundancy is after all the essence of multi-tiered rescue response systems.

If the IC decides to cancel assisting units before the rescue has been completed and before rescuers have left the high-risk zone, he subjects innocent victims to additional minutes (and in some cases, hours) of pain, fear, and medical risk. ICs also risk losing firefighters and other rescuers to preventable mishaps and avoidable consequences when mishaps occur.

We can argue about why an otherwise competent IC would conduct a technical rescue operation in such a manner, clearly ignoring the best interests of a trapped patient and possibly the safety of his own personnel. The many possible reasons include inexperience, lack of training about the hazards at hand, an inaccurate size-up of the conditions, or poor advice from supposed *authorities*. Sometimes the answer is simple overconfidence or even personal pride that overrules common sense and concern for the welfare of the trapped victim and compels an IC to take whatever measures are necessary to ensure that their company—and not the rescue company—frees the victim. But one thing is undeniable; this kind of decision is made by ICs more frequently than is acceptable.

As just one of numerous examples, this author recalls a situation in which units were responding to an emergency involving a young girl trapped beneath a hinged storm drain-cover that weighed in excess of 400 lb. The girl was lying in several inches of water on the bottom of a flood-control channel with the edge of the hinged cover pinning her hips. The victim and some friends had been swinging the cover back and forth, timing their ranger crawls to gain entry into an underground section of the drain. Unfortunately, the victim had been caught by the cover as she attempted to enter.

The captain of the first-due truck company was so certain that his crew could pry the cover off the girl with a standard set of hydraulic spreaders that he advised the IC to cancel the rescue company response. An engine company officer, who was running the incident until the arrival of the battalion chief, canceled the additional response even before the truck company

had a chance to remove the hydraulic tools from its truck. The rescue company captain decided to continue driving to the incident *non-emergency* to observe the operation for training purposes. He arrived 25 minutes later to find that the truck company had been unable to gain a hold that would allow them to budge the metal cover with the hydraulic spreader, and now they were attempting to use a rescue saw to cut the drain's hinges that threatened to crush the girl.

Fortunately, the battalion chief also arrived on the scene and shared the rescue captain's concern about the operation. The girl was becoming hypothermic from the water swirling around her, and it was obvious that the truck company captain had erred, and not on the side of the patient's safety. The battalion chief took over command of the incident, halted the truck company's unsafe operation, and assigned the rescue company captain as rescue group leader. Within ten minutes the cover had been properly cribbed and supported, rescue air bags had been placed on either side of the girl, and the cover had been carefully lifted sufficiently high to allow her to be extracted and removed from the water on a backboard. The moral of this story is not about the capabilities of the rescue company; rather, it is about the truck company captain's apparent determination to conduct the rescue to the detriment of an innocent victim who relied on the firefighters to act in her best interests.

Rescue rule 4: always assume it's a rescue until evidence proves it's a recovery

Never assume victims are beyond help until every indication demonstrates that there is no chance of rescue. How many times have we heard of situations where fire/rescue professionals arrived on the scene of a rescue emergency and pronounced that the situation was a *recovery*, only to discover minutes or hours later that the victim was in fact alive and waiting for help? Figure 6–36 shows some rescuers conducting *selective debris removal* in the collapse of the Pentagon on the fourth day after the 9-11 attacks. At the same time, FEMA USAR task forces work to stabilize and search other areas of the building not visible in this photo. Also unseen in this photo are the many *potentially survivable void spaces* that honeycombed the pancake collapse area. Each of these spaces held the potential for finding one or more live victims, and a physical search of each space was required to determine beyond a reasonable doubt that no survivors were left waiting for help. This emphasizes the point that rescuers should assume that there may be survivors until strong evidence demonstrates otherwise.

Fig. 6–36 Selective Debris Removal

This author has, upon arrival at the scene of several complex rescue emergencies, been told by ICs or other officers that "it looks like this is a *body recovery*" because victims were missing or not obviously visible. Later, during the course of a proper search and reconnaissance operation, it was discovered that there indeed were live victims awaiting rescue.

One time, upon arriving and looking into a deep collapsed excavation to see the hard hat and one hand of a construction worker buried past his eyes in compact soil, this author whispered to a firefighter already on the scene, "This doesn't look too good."

I was then surprised to hear an L.A. city firefighter respond, "Yeah, but he's talking to us through the dirt." I had wrongly made the assumption that the worker must surely have succumbed to the soil that appeared to be tightly packed around his entire body and face. However, the worker was indeed buried alive, and after an intense and dramatic eight-hour operation by city and county firefighters, he was successfully rescued. The man eventually returned to his family with little or no physical deficit. This event was aired live across the United States on television.

In Figure 6–37, L.A. City and County firefighters work to rescue a man buried alive by an excavation collapse. This rescue, aired live on national television until its conclusion eight hours later, was a textbook case of rescuers working full-blast to locate and extract a victim whose fate hung in the balance and who was actually buried and reburied by several secondary collapses (including one 30-minute period where he was entirely covered by a thick layer of soil).

Fig. 6–37 Excavation Collapse (Courtesy Mike Meadows)

Some observers described it as a miracle rescue. The man was fortunate that first-arriving firefighters had enough training and experience to realize that they should attempt making contact with him to ascertain his true status. He was also the beneficiary of a *no holds barred* rescue operation that eventually included the response of more than 100 firefighters from two fire departments. The response team included several soils engineers, some tunneling experts, members of two FEMA USAR tasks forces who happened to be on duty, three hydro-vac truck units, an emergency room physician, a paramedic team, and many other resources.

Rather than a miracle, the man's survival could also be attributed to good training, an effective multi-tiered USAR response system, and multi-agency cooperation and coordination. The scene also included the proper equipment, highly experienced USAR-firefighters, an IC who was willing to employ all available resources to get the job done right, and the perseverance of firefighters and rescuers who simply would not give up until the victim was extracted and brought to safety.

Rescue rule 5: if something tells you there is a missing or trapped victim somewhere, conduct a thorough search before declaring "all clear"

Some rescue emergencies have a tendency to trick ICs into thinking that there may be no victim when in fact a patient is trapped somewhere out of sight and in dire need of help. An example of this is a vehicle over-the-side of a mountain road that appears to be unoccupied (e.g. a stolen push-over, an abandoned vehicle, etc.). Another example is a 911 call reporting an aircraft down in mountainous terrain with no obvious sign of a plane crash and no emergency locator transmitter (ELT) signal. Other examples include unconfirmed radio reports of boats sinking or in distress but no sign of a craft on the surface, an unconfirmed cell phone call from a purported driver who has driven over-the-side with no evidence of a car crash, or a building collapse with no sign of victims. Still other cases may involve avalanches over a road with no obvious victims, trench collapses without a proper accounting of all workers, and structures collapsing without an accurate record of all the occupants inside at the time (Fig. 6–38).

Fig. 6–38 Rescuer Assessing the Progress of Void Space Search and Selective Debris Removal Operations at the 9-11 Pentagon Attack Collapse (Courtesy Ruben Almaguer)

In Figure 6–39, L.A. County firefighters use griphoists and rope systems to roll a car that pinned its driver after going over-the-side on a mountain road. The unfortunate driver was pinned at the hips on a rock and was burned alive while passersby attempted to assist.

Fig. 6–39 Operational Briefing to Clarify Information

This author has arrived at the scene of reported vehicles over-the-side and been told by the IC that there were no victims, when in fact the first responders had yet to reach the vehicle. First responders had not checked the trunk for potential victims of crime,[5] thoroughly searched the chaparral, or conducted a helicopter search from the air for victims who might otherwise not be seen. The IC apparently hadn't been trained to know that it's possible to spot ejected victims by searching with thermal imagers and night-vision systems from the edge of the road. This can be a serious oversight if in fact a live victim is in need of timely assistance yet hidden from plain view.

This brings up another important point about managing unusual rescue situations. It is part of the role of rescue and USAR unit personnel to act as technical advisors to ICs and officers who may have less experience managing technical SAR operations. Sometimes experienced rescuers will identify telltale signs of survivors that have escaped the attention of first responders and the IC. This is not an indictment of the first responders or the IC by any means. Rather, it is simply a recognition that first-arriving captains or ICs may never have seen an incident of this type before and are unaccustomed to sizing it up, evaluating the potential for missing victims (and their likely locations), and coordinating an integrated SAR operation using ground-based or aerial resources.

If the first responders have never seen a particular type of rescue emergency at any time in their careers, it's impractical to expect them to operate as if they do it every day.

This is one reason why the fire service is embracing the concept of highly trained, highly skilled, and highly experienced firefighters/officers assigned to dedicated rescue companies and USAR units that specialize in identifying, assessing, and managing unusual rescue emergencies. History has demonstrated that fire/rescue agencies that employ this approach have a measurably higher rate of success in managing unusual and highly dangerous or complex SAR-related emergencies and disasters. Time and again, the intuition of the trained rescuers, honed by their schooling, study, and experience, has proved effective in locating and rescuing victims who might otherwise have gone undetected (and therefore would not have survived).

For example, in 1988 a train carrying hundreds of tons of a powder known as *trona* derailed at the bottom of a steep downgrade in southern California. Nearly a dozen homes were buried by the debris and trona in the morning-hour crash. After several hours of physical searches and attempts to identify who was home at the time of the crash, local firefighters and officers came to the conclusion that there were no more victims, even though several homes remained buried roof-high in the powder.

Fortunately, Deputy Chief Mark Ghillarducci arrived on the scene to address coordination issues for which his agency was responsible during state disasters. Ghillarducci had a hunch that there might be one or more uncounted victims buried somewhere in the morass of the crash site. He was concerned that no canine search teams had been sent to comb the area for signs of trapped victims. Ghillarducci was successful in convincing the local IC that the scene couldn't be declared clear of victims until canines had searched the area and until all potential survivable void spaces had been physically searched.

With approval of the IC, Ghillarducci arranged for two canine search teams to be flown from San Francisco to the scene. In those early days of USAR, there were few certified USAR search canine teams in California, and even fewer in most other states. Sure enough, the dogs *alerted* on one of the homes. Firefighters began tunneling through the roof of the home, which had been partially crushed by a freight car. To their surprise, they established voice contact with a man who was pinned beneath a truck that had come through the roof of his home as he lay on a couch watching television nearly nine hours earlier. An intense seven-hour rescue effort ensued, and the man was pulled to freedom about 16 hours after he had been trapped.

Had it not been for the intuition of an experienced rescuer who followed his experience-honed instincts, and had it not been for the willingness of the IC to follow his recommendations, the live victim may have become a deceased one. A life was certainly saved that day, and it stands as one of countless examples of rescuers following their common sense and intuition to a successful conclusion.

Fifteen years later a similar train derailment occurred in Commerce (L.A. County), and it was evident that the lessons learned in the 1988 trona train wreck had been learned and institutionalized. In June 2003, 31 rail cars escaped a train yard and headed downhill toward downtown Los Angeles, quickly outpacing a desperate attempt by railroad personnel who gave chase in a locomotive. The train raced through neighborhoods at speeds approaching 87 miles per hour, heading for a rail yard where cars laden with propane and innumerable hazardous materials were in the way. Thirty minutes after the cars escaped the rail yard, frantic rail employees made a fateful decision to derail the train in a mixed residential and commercial area near the heart of Los Angeles. The railroad did not notify law enforcement, the fire department, or residents that a runaway train was speeding toward the city. The resulting derailment crushed homes beneath hundreds of tons of debris and rail cars and resulted in a major collapse SAR operation that lasted two days.

The LACoFD dispatched a first-alarm *derailment* response that consisted of four engines, one truck company, a paramedic squad, one hazmat task force, one USAR task force, one foam unit, and two battalion chiefs. First-due engine 22's captain could see a cloud of dust rising in the distance, and as he got closer he could see a cluster of rail cars rising 30 feet above homes for at least 200 yards. It was apparent that some homes had been crushed. He immediately called for a second alarm, which was quickly upgraded to a third alarm with three fire/rescue helicopters, a dozen ambulances and paramedic squads, and literally dozens of additional resources.

Among the most critical priorities was to conduct a thorough search for trapped and missing victims. Assistant Chief Mike Bryant became the IC; he quickly ordered responding USAR resources to report directly to the command post to quickly establish a massive canine and technical search to back up the first responders who were at that time conducting primary searches.

At least five FEMA/California OES-certified USAR canine search teams were dispatched from LACoFD and surrounding departments, and canine search operations were conducted at various times for the next 30 hours. Technical search operations led by personnel from two LACoFD USAR task force stations were in progress for more than a day; fire department units were on the scene for several days. Amazingly, considering the speed of the train at the time of derailment, there were no fatalities in this wreck and only 12 people were injured.

The low casualties were due in part to the mid-afternoon time of the derailment while school was in progress and many adults were at work. But the massive and sustained response of firefighting, EMS, and USAR resources (including USAR canine search teams) was indicative that the lessons from the trona train wreck had indeed been heeded, which leads to rescue rule 6.

Rescue rule 6: learn from the mistakes and achievements of others.

We know that one of the most effective ways to learn is by observing and studying the efforts of others, carefully dissecting the conditions that lead to either success or failure. This can be done directly by physically observing ongoing rescue operations, or indirectly, such as watching on television, videotape, or some other medium. It can be accomplished through case studies, by reading articles in trade magazines and newspapers, by discussing the rescue with those who were involved, and other methods.

This kind of study should not be limited to your own city, county, state, or nation. Every year there are worldwide examples of rescue emergencies and disasters that should catch our attention and are worthy of our examination. Those who do not study the failures of others may be doomed to repeat their mistakes. Those who do not study the successes of others may miss the chance to replicate their triumphs by capitalizing on proven methods and approaches.

Rescue rule 7: learn from your own mistakes and successes

This rule should also be obvious, but it's sometimes the most difficult to follow, particularly when an incident goes bad and mistakes are made. Certainly, we want to capture the lessons we learn through making successful rescues and should be willing to pass on those lessons to others within our own units, agencies, states, and even beyond.

The hard part is when mistakes are made during problematic rescue emergencies. There is a tendency in many agencies to gloss over failures, particularly when they resulted in death or suffering of victims or rescuers and when there is a potential for civil litigation or even criminal action if word gets out that things could have gone better.

While there are legal and public relations sensitivities that must be recognized, there are constructive ways to capture, learn, and pass on lessons from problematic rescue operations. This can take the form of informal discussions among those involved in the operation, agency-specific post-incident case studies, and even published studies that recognize both the positive and negative aspects of a particular operation. *Coming clean* is sometimes the best way to reduce the impact of a problematic emergency operation because criticism from the public, the press, and public officials may be doubled if there is an appearance that the agencies involved attempted to squelch information and prevent rescuers from improving the next operation by sharing the information in a usable form. It starts with the willingness of every IC, every officer, and every firefighter/rescuer to be honest with himself about what happened, what was done right, and what could have been done better. It extends to an agency philosophy and culture that encourages the free and honest exchange of information, particularly when such information can be used to improve the life safety of citizens and rescuers in the future.

Endnotes

1   Since the mid 1980s, the Dade County and Fairfax County USAR task forces have been dispatched to disasters around the world under the auspices of the USAID and OFDA. USAID/OFDA coordinates the operations of American USAR assets during disasters on foreign soil.

2   In some departments, certified safety officers are dispatched to assume this responsibility during working fires or multiple-alarm incidents.

3   Conversely, in some jurisdictions, USAR incidents may be a regular occurrence, and certain types of fires may be the rarity.

4   "OSHA: Violations Not Cause of Fire Deaths." *The Worldlink*.com (May 21, 2003).

5   There have been a number of suicides, homicides, and other situations across the United States where victims both live and dead have ended up in the trunks of cars that were pushed over cliffs or over the side of mountain roads. In fact, it's such a common modus operandi of homicide in some places that rescuers should make a special point of checking the trunks of vehicles discovered over-the-side in mountainous areas and where roads are close to cliffs.

Appendix I: Sample SOG for Mud/Debris Flow Rescue Ops

The following are examples of SOGs developed for a specific rescue hazard; in this case, mud and debris flow emergency operations. This sample may be used as a template for other SOGs, or it may be adopted by fire/rescue agencies that have mud and debris flow hazards.

Mud and Debris Flow Emergency Operations

I. Introduction

A. *Purpose*: To ensure the ability to manage the consequences of mud and debris flows and other related events in a safe and effective manner.

B. *Scope*: This instruction applies to all personnel at the scene of a mud and debris flow emergency.

C. *Author*: The _____ shall be responsible for content, revision, and annual review of this instruction.

D. *Background*: Mud and debris flows are among the most hazardous situations because they combine some of the most dangerous characteristics of flash floods and mud slides. They have been known to wipe out entire cities and kill tens of thousands of people. They are preceded by little warning (like flash floods) and they can leave cities buried under dozens of feet of mud and rocks.

Worse is the fact that mud and debris flows have a tendency to flow in waves or surges. In other words, personnel attempting to rescue people from a mud and debris flow may be in imminent danger of progressively larger waves or surges of mud, water, and debris that can strike with little warning. Therefore, the IC must be mindful of the danger to any personnel who are committed into the danger zone, and should take appropriate precautions to protect rescuers and victims.

Mud and debris flows are endemic to Los Angeles County, where steep mountains and foothills rise above valleys and flood plains. They are especially common in parts of the county where the mountains have been mobilized and fractured by local tectonic forces, creating steep, unstable slopes that are vulnerable to erosion. Because the local mountains are covered with highly flammable vegetation that is burned at regular intervals, the probability of mud and debris flows is especially high because vegetation is sometimes the only thing keeping boulders and soil clinging to the slopes. When fire denudes the vegetation, the rock and soil begin sliding and falling into the canyon bottoms. When intense rain occurs, tremendous amounts of debris can be quickly turned to slurry and mobilized into a large flood of mud, rock, and water.

II. Responsibility

A. All ICS shall be responsible for observing the safety concerns addressed in this instruction.

B. All personnel shall be familiar with these instructions.

III. Policy

A. Upon arrival, the IC shall make a comprehensive size-up.

B. The IC shall address the following immediate concerns:

1. Assess the condition of the incident site, including the stability of slopes and hazards such as deep or moving mud and debris, shifting structures, etc.

2. Determine the potential for secondary mud and debris flows or mud slides.

3. Assign an upstream safety to provide constant observation for potential hazards.

4. Determine the number and location of victims.

5. Begin stabilization in preparation for rescue operations.

6. Begin rescue operations.

7. Coordinate pre-hospital care and transportation.

IV. Procedures

A. Upon arrival, the IC shall conduct a size-up that includes the following:

1. Static or dynamic situation? Is the mud and debris still flowing, or has it come to rest?

2. Approximate number of victims trapped and/or missing.

3. Any structure collapses that are evident.

4. Any structures or vehicles in danger of secondary mud and debris flows.

5. Safe approach for other responding resources.

B. The IC shall address the following immediate concerns:

1. Assess the condition of the incident site, including the stability of slopes and hazards such as deep or moving mud and debris, shifting structures, etc.

   a) The IC may request the response of professional county, city, and/or private geologists and other experts who can provide accurate assessment of site conditions.

   b) Departmental USAR personnel may be assigned as technical advisors and other appropriate positions to ensure accurate assessments.

2. The potential for secondary mud and debris flows.

   a) Assign an assessment team to check for landslides that are blocking streams and for signs of impending mud and debris flows, etc.

b) Assign an upstream safety officer to provide constant observation for secondary surges of mud and debris, secondary mudslides or landslides, flash floods, etc. The upstream safety should be in a position to visualize the potential hazards. This may require that they be positioned on mountain peaks, on high roads, or even in a helicopter over the incident site. The upstream safety should have multiple warning methods (i.e. radio, whistle, air horn, runners, etc.) to alert downstream personnel of a hazardous situation.

3. For potential slope slippage, the IC may request geologists and other experts who can assist the upstream safety by providing instrument monitoring to determine whether any movement is occurring and to warn of hazardous conditions.

4. Determine the number and location of victims.

   a) Assign search teams, search group, canine search teams, technical search teams, aerial search teams (departmental helicopters) as conditions dictate.

5. Begin stabilization in preparation for rescue operations. Depending on conditions, this may involve one or more of the following:

   a) Establishing high-line rope systems and other high-angle methods.

   b) Shoring-up unstable slopes and structures.

c) Mitigation of electrical and other infrastructure hazards.

d) Clearing access roads and safe refuge areas with heavy equipment.

e) Draining water and mud from the site.

f) Creating diversion dikes and dams, etc.

6. Begin rescue operations. Depending on conditions, this may involve one or more of the following:

   a) The use of heavy equipment and helicopters to transport personnel, equipment, and victims in and out of the incident site.

   b) Tunneling through the roofs and sides of buildings, and tunneling through debris to reach victims.

   c) Using IRBs attached to rope systems to access victims.

   d) Helicopter hoist operations.

   e) Lifting and moving heavy objects.

   f) Cutting and moving concrete slabs.

7. Coordinate pre-hospital care and transportation.

Depending on conditions, this may require the use of heavy equipment and helicopters to move patients and personnel, mass casualty incident operations, special treatment modes for hypothermia, crush syndrome, compartment syndrome, near-drowning, etc.

Sample Response Matrix for Mud and Debris Flow Emergencies

First Alarm	Second Alarm	Third Alarm
4 Engines	4 Engines	4 Engines
1 Truck	2 Trucks	1 Truck
1 USAR task force	2 Squads	1 Squad
1 Squad	2 Camp crews	2 Camp crews
1 HMTF	2 USAR trailers	1 Sup.
1 BC	2 BC	2 BC
	1 AC	1 AC
	1 Air squad	1 DC
	1 USAR task force	1 USAR task force

Table 6–1 Sample Response Matrix for Collapse Structure

First Alarm	Second Alarm	Third Alarm
1 Swift-water team	1 Swift-water team	1 Swift-water team
4 Engines	4 Engines	4 Engines
1 Truck	2 Trucks	1 Truck
1 USAR task force	2 Squads	1 Squad
1 Squad	2 Camp crews	2 Camp crews
1 HMTF	2 USAR trailers	1 Sup.
1 BC	2 BC	2 BC
	1 AC	1 AC
	1 Air squad	1 DC
	1 USAR task force	1 USAR task force

Table 6–2 Sample Response Matrix for Mudslide/Debris Flow

Appendix II: Notes on Command of Urban Search and Technical Rescue Incidents

I. Use LCES

II. Risk management is controlling the degree of risk to rescuers

III. Important personal characteristics of rescue commanders include the following:

A. Competence in the position

B. Competence to evaluate conditions and develop effective strategy

C. Can manage a balance of aggressiveness and safety

D. Is confident enough to seek bad news and others' ideas

E. Maintains calm and rationality under duress

F. Resists stampede mentality

G. Is flexible enough to reverse decisions when necessary

H. Has ability to make own decisions and challenge bad ones

IV. Command process:

A. Rule out no win options

B. Get the following essential information:

1. Hazards

2. Situation

3. Objectives

4. Strategy and tactics

5. Constraints

6. Communications nets

7. Organization of incident

8. Resources

C. Establish priorities

1. Rescuer survival

a) LCES

b) Awareness (situational)

c) Clear, calm thinking

d) Risk Management Process

2. Incident objectives

3. Tactical objectives

D. Establish realistic expectations

E. Seek and stay current with all serious risk

  1. Hazard assessment

    a) Potential behavior of damaged buildings, etc.

    b) Identify hazards as they occur

    c) Analyze each new hazard

    d) Tactical decision

    e) Discussion

  2. Risk control

    a) Rank degree of risk

    b) Discussion

    c) Determine necessary risk vs. gain

    d) Mitigate risk:

      1) Better LCES

      2) Change tactics

      3) Operational retreat

      4) Monitor and attempt later

      5) High-risk tactic guidelines

      6) Rapid intervention

  3. Enforcement:

    a) Accountability

    b) Rapid Intervention

    c) Communication

    d) High Risk Tactics

    e) LCES

  4. Risk management process

---

[1] Provided by Battalion Chief (retired) Gary Nelson, LACoFD.

Appendix III: Rapid Intervention During Rescue Operations

Contrary to popular belief, RIC operations are not limited to firefighting mishaps. In fact, ICs of non-fireground emergencies are negligent if they do not take into account the potential need for RIC operations during the course of non-fireground emergencies.

Fortunately, most progressive fire departments understand the need for RIC capabilities to be applied to every emergency involving high life hazards, where a mistake or mishap may result in firefighters or other rescuers becoming lost, trapped, or injured.

Figure 6–40 shows a seven-hour rescue in the San Gabriel Mountains using a two-person entry team of firefighters from an LACoFD USAR company (right, wearing SCBA). The firefighters are consulting with support personnel. They will be backed up by a two-person RIC (left, wearing SCBA) as they make their way into an abandoned gold mine to assess the predicament of a man trapped in a 900-ft vertical shaft.

Fig. 6–40 Mine Rescue in the San Gabriel Mountains (Courtesy Gary Thornhill)

Emergencies typically requiring RIC capabilities include:

- Fireground operations
- Confined-space rescue
- High-angle rescue
- Collapse SAR

- Dive rescue

- Marine disaster/marine rescue

- Avalanche rescue

- Landslides and mudslides

- Mud and debris flows

- Flood rescue

- Swift-water rescue

- Helicopter-based technical rescue

- Deep shaft rescue

- Other high-risk SAR operations

On arrival

Figure 6–41 shows an LACoFD USAR company arriving on the access road to the abandoned gold mine. Because the road was steep, muddy, and required 4-wheel drive, all of the company's confined space/tunnel rescue equipment was offloaded and transferred to 4-wheel drive brush fire patrols, which shuttled it to the forward equipment staging area.

Fig. 6-41 USAR Company Members Assessing Mine Rescue Problems Prior to Entry

- Apparatus placement for best access to RIC equipment (avoid collapse zone). While RIC leader reports to IC, RIC members establish equipment pool.

- RIC leader reports to incident command for briefing and review of:

  - Current situation/status (what is happening, where are victims, what is their predicament, where are units and personnel)

  - Operational mode (offensive, defensive, rescue, etc.)

  - Personnel accountability

  - RIC frequency and contingencies

- RIC leader briefs RIC and together they evaluate the incident site

  - Take a *hot lap* around the scene

  - Evaluate structural integrity

  - Access points

  - Make additional access points as necessary. Make them large, because they are also exit points.

  - Throw additional ladders (or request additional ladders to be placed)

- Evaluate incident-specific hazards

- Observe effectiveness of current tactics and strategy to anticipate possible problems and RIC situations

- If offensive fireground operations are occurring, evaluate interior and exterior

- If the fireground situation is defensive, evaluate exterior

- If a technical rescue, evaluate the actual rescue site and surrounding areas

- Observe for unusual signs (e.g., terrorism, secondary devices, secondary collapse potential, additional flood surges, aftershocks, etc.)

- Discuss contingencies

In Figure 6–42, a confined-space trained Haz Mat Captain conducts an operational briefing prior to a confined space entry operation.

Fig. 6–42 Briefing

Fig. 6–43 Preparing for Rapid Intervention Operations

- Remember, no side jobs or freelancing. If you're assigned as RIC, you and your crew should perform no other task that prevents mounting an immediate rapid intervention operation. The IC should request sufficient resources to address other fireground and/or rescue needs.

- Review the RIC equipment pool to ensure that it is appropriate for the hazards at hand.

  – Fireground RIC equipment

    ○ *Per each member.* Two or three light sources, personal tools, (1-hour bottle if available, or consider closed-circuit systems), drop bag, webbing, rapid intervention strap, or other capture and harnessing device

    ○ *For RIC.* Thermal imager(s), extra SCBA with mask or SABA for trapped firefighter/rescuer, rotary rescue saw, sawzall with extra blades and batteries (or electric), portable battery-powered hydraulic rescue tools, air bags in bag with regulators and hose, SCBA bottle for air bags, hydraulic jack, hydraulic or pneumatic emergency shores, attic ladder, debris bag with handles to carry out firefighter/rescuer, axes and belts, sledge hammer, rabbit tool, irons, sounding tool, chain saw, and other tools based on hazards at the scene

      - Rescue RIC equipment pool (Fig. 6–43).

    ○ *Per RIC member.* PPE, three sources of light, respiratory protection based on hazards, rescue harness for high-angle situations, and other equipment based on hazards.

    ○ *RIC team.* Rescue equipment based on hazards at the rescue scene.

      - *Confined-space rescue:* SABA or bio-packs for victims as well as rescuers. Intrinsically-safe monitors, three sources of light, and other standard tools for confined-space entry. Air-powered tools for extrication, confined-space litters and rescue harnesses, rapid intervention straps, and high-angle equipment.

      - *Water rescue:* Throw bags and high-angle gear, IRBs or PWC, helicopter-based swift-water teams, or helicopters equipped with cinch harnesses and other self-rescue devices that can be deployed from a copter. Whistles, bullhorns, radios, and other signaling devices.

      - *Trench/excavation rescue:* Air knives and air vacuums, hydro-vac trucks, ladders and other escape devices, digging tools, rapid shoring devices, and high-angle gear.

RIC standby during incident

USAR firefighters assigned as an RIC stand by during a technical rescue operation and:

- Monitor incident frequencies

- Observe fire behavior and other factors

- Observe effects of tactics and strategy

- Watch for signs of impending collapse, back draft, etc.

- Watch for rescue-related hazards such as secondary collapse, flood surges, rope operation problems, helicopter mishaps, etc.

- Think LCES

RIC situation occurs

USAR firefighters from the L.A. County FD and the Orange County (CA) Fire Authority work in the danger zone to extract the victim of a fatal excavation collapse. The shoring is intended to prevent a 13-ft excavation wall from suffering a secondary collapse that might bury rescuers. In the background, an RIC consisting of USAR-trained personnel is on standby.

- Assess the event with IC. What just happened? To whom did it happen? Where were they last seen? Are you in contact with them? What is likely to happen next? What is your best strategy to get to the victim(s) and extract them without delay?

- Assemble RIC and equipment; give quick briefing and move in.

- Ensure use of RIC frequency.

- Monitor other frequencies to communicate with trapped/lost personnel, to maintain awareness of other conditions, and to request assistance as necessary.

- Make access to victim or to search area.

- Assess situation and take appropriate action (search, extrication, etc.).

Appendix IV: Sample Technical Rescue SOGs (General)

The following is a sample of SOGs for a variety of technical rescue emergencies. They may be adopted, adapted, or used as a template for rescue SOGs.

Technical Rescue/USAR Incidents (General)

The USAR program is designed to provide a timely, effective response to all technical rescues and to ensure firefighter and rescuer safety. First responders may not always recognize hazardous conditions or the need to use special equipment, strategies, and methods to safely manage the incident. It is the job of personnel assigned to USAR companies to provide technical support, special equipment, and trained personnel to get the job done and maintain reasonable safety at the incident. It is the responsibility of the IC to make the best use of USAR resources to ensure the best outcome for victims and rescuers.

En route

Do not cancel USAR resources until the victim has been extracted, and until *all* personnel are out of potential danger zones (i.e. trenches, collapsed buildings, over-the-sides, confined spaces, etc.). As long as rescuers and victims are in the hazard zone, the IC should continue the USAR resources to provide technical advice, to act as trained safety officers, to provide backup rescue capabilities, and to assist with the rescue as needed.

- Based on the incident type, the incident text, and other information, consider the potential need for additional resources.

- As necessary, consult with USAR company personnel to determine the need for special resources.

- Consider the response time of USAR companies to your incident. In some cases, it will help to expedite their response via helicopters.

- Consider the need for air squad hoist operations and other air operations capabilities, and request them early in the incident.

At the scene

Be prepared for long-term operations. Because of unforeseen complications, many technical rescues take twice as long to safely complete than initially anticipated. Consider rotating crews and providing rehab to reduce fatigue and maintain good working strength until the rescue is accomplished.

- Perform size-up to determine special hazards and needs.

- Develop a strategy to perform the rescue if on-scene personnel have the proper training and equipment to conduct the operation safely and effectively.

- If the incident exceeds the capabilities of first responders, develop a strategy to protect the patient, protect rescuers, and stabilize the incident until the arrival of USAR units. If necessary, isolate the danger zone and deny entry until USAR units arrive. Consider the need to provide indirect assistance to the patient (i.e., lowering an oxygen mask and helmet to the patient, providing ventilation, preventing secondary collapse, etc.).

- Consult with USAR companies regarding special needs.

- Request additional resources early to avoid critical delays.

- Consider the need for unified command on large or complex incidents.

High-angle Rescue Operations

Size-up

- Exact location of incident and access for ground units and helicopters.

- What is victim's predicament (trapped or stranded on cliff, fallen to bottom, etc.)?

- Is victim able to assist in his/her rescue (injuries, age, frozen in place, etc.)?

- Is this a situation best handled by a safe helicopter hoist operation, a ground haul rope rescue, or a combination of both?

- For vehicles over-the-side, is there an extrication problem? What equipment is needed?

Resource allocation

Rope rescue situations are generally dispatched as one of the following response types:

- *Traffic collision over-the-side (TCO)*. One each: engine, paramedic squad, BC, USAR company, truck company, and helicopter.

- *Cliff/coastal*. Two each: engines and lifeguard bay watches and lifeguard rescues (for coastal incidents); one each: squad, BC, USAR company, and helicopter.

- *Person trapped*. Two engines; one each: truck company, USAR company, squad, and BC.

- *Confined-space rescue*. Two engines; one each: squad, BC, USAR company, hazmat task force, and mobile air unit.

If the USAR company is more than 30 minutes away, consider helicopter transportation of USAR company personnel.

Assignments

For most standard rope rescue operations, the following positions can be staffed by first responders. These positions may, at the IC's discretion, be assigned to USAR task force personnel and other USAR technicians.

- One top division supervisor to supervise top division rescue operations

- One bottom division supervisor to supervise bottom division rescue operations

- One safety officer to double-check all systems and help ensure overall safety

- One rigger to establish and oversee the operation of technical rope systems

- One or two primary rescuers to make contact with the victim, perform a cliff pick-off if necessary, protect the victim from additional falls and rocks, and properly package the patient in a rescue litter if necessary.

The following additional positions are required for most rope rescue operations. These positions may be filled with first responder personnel, preferably with oversight from USAR personnel:

- One main line tender

- Two or more personnel as the lowering and/or haul team on the main line

- Two litter tenders (if needed to augment the original rescuers)

- One or more observers (where needed to maintain constant visual contact with the rescuers during raising and lowering)

- If the air squad is not utilized for a helicopter hoist evolution, the IC may request that the air squad land and assist with establishing and operating rope rescue systems, treating and packaging the patient, or other essential tasks. The air squad may be used to fly the patient to a trauma center after the patient is extracted.

- It is strongly recommended that the IC consider requesting a camp crew to clear a path through brush to help carry patients in the backcountry and to staff haul teams.

Scene control

ICs should not cancel USAR resources until the victim is rescued and until all rescuers are out of the danger zone. Canceling USAR resources prior to this benchmark may delay the rescue of the victim and may create unnecessary risks to fire department personnel.

Rescue operations

- Determine whether a helicopter for hoist evolution will be advantageous to extract the victim or to deploy rescuers to the victim's location. A risk vs. gain analysis should be conducted by the IC with input from the air squad crew and USAR-trained personnel to help determine whether it is best to conduct a helicopter hoist operation, ground-haul rope evolution, or a combination of both to extract the victim(s).

- If rope systems are used, establish *bomb proof* anchors. If vehicles are used as an anchor, chalk the wheels and remove the keys from the ignition.

- Any point where a rope passes a stationary object must be padded with edge protection to prevent catastrophic damage to the rope.

- Gloves, helmets, and other PPE shall be worn by all members involved in the operation. At night, headlamps are recommended, and all members should carry at least one (preferably two) backup source(s) of light. Goggles are mandatory around helicopters. Only NFPA-approved harnesses or other approved rescue harnesses are permitted. Personnel should not wear recreational climbing harnesses for rescue work.

- A helmet and goggles should also be provided for the victim and can be taken to the victim by the first rescuer or by the litter team.

- If there is only one rescue rope on scene (i.e. patrol on scene first), the first rescuer may rappel on a single line with a three-wrap Prusik for fall protection. The first rescuer should bottom-belay other rescuers who rappel on the single line.

- When feasible, two ropes should be used to lower the first rescuer over the side and always when raising victims and rescuers. The second rope will be the safety line.

- In some cases (e.g. vertical cliffs where the rescuer will be out of sight and requires split-second control to reach the victim), the rescuer may rappel on the main line with a safety line attached to his sit harness and chest harness for fall protection.

- Non-rescue-rated cable winches shouldn't be used to raise victims.

- Even rescue-rated cable winches should be used judiciously. Caution is advised whenever the cable winch operator does not have direct line-of-sight visual contact with the rescuers. In the event of communication delay, it is possible for the powerful cable winch to pull rescuers through brush and rocks before the operator realizes it.

- Injured victims should be securely packaged in a rescue litter before any raising or lowering

operations. Ferno-Washington litters should not be used for helicopter hoist operations because of their aerodynamic properties beneath rotor wash.

- If a helicopter hoist evolution is selected to extract the victim, at least one air squad crewperson (or a USAR company member) will be inserted to the victim's location to ensure proper packaging of the patient before hoisting operations commence. Generally, a pre-rig from the air squad will be used for hoisting. Air operations pre-rigs differ slightly from standard departmental pre-rigs to allow the litter to be inserted more readily into the cabin of the copter. However, standard departmental pre-rigs can be used if they are in the full-cinch position. The air squad crewperson or USAR task force members will double-check the pre-rig prior to hoisting.

- When the cable is lowered from the hovering air squad, the air squad crewperson or a USAR company member will connect the cable to the steel ring of the litter's pre-rig, and this member will manage the tag line as the litter is raised to the copter. The purpose is to avoid complications during the critical moments when the helicopter is hovering overhead, when radio communication is difficult, and when the copter is connected to a litter with a tag line extending to the ground with the potential for entanglement.

Marine Emergencies

The marine emergency system is activated for incidents where victims are in need of rescue on the open sea (or in some cases, lakes) resulting from airplane crashes, capsized boats, boat collisions, and boat fires, etc.

When the marine emergency plan is implemented for a marine disaster, the fire department will automatically respond with other agencies in a coordinated response of marine-based, air-based, and land-based fire/rescue/EMS resources. They will effect timely rescue and recovery, provide rapid transportation to shore, and process (as necessary) through a landslide multi-casualty incident system.

Ultimate responsibility for offshore SAR operations along the coast belongs to the local U.S. Coast Guard district's federal SAR coordinator (SMC). However, the first on-scene agency will establish command. Upon arrival of the Coast Guard and other agencies, unified command will generally be utilized.

Resource allocation

The first-alarm assignment for marine emergency incidents consists of the following:

- Four engines (two for landslide operations)

- One truck for landslide operations

- Four bay watches

- Three lifeguard rescues

- Two lifeguard area supervisors

- Two fire boats

- Three air squads

- Four squads (three for landslide operations)

- Two USAR companies

- Three battalion chiefs

- Personnel regularly assigned to USAR companies are trained as surface rescue swimmers, capable of deployment from helicopters or boats.

- Lifeguard division rescue swimmers and divers may be deployed from bay watch boats and IRBs.

- The U.S. Coast Guard may dispatch rescue helicopters and cutters to the scene.

- Other fire departments may dispatch ground units to assist with multi-casualty incident operations.

- The IC may request the response of large hovercraft(s) from the U.S. Navy, with an ETA of approximately 1 to 2 hours. The department and the Navy have an agreement to use hovercraft to transport engine companies, rescue units, and special equipment to support emergency operations.

Weather, helicopter availability, and other conditions permitting, USAR companies shall respond directly to the scene as a helicopter-based rescue swimmer team with ten-person inflatable rescue platforms to begin removing victims from the water. If conditions preclude helicopter deployment, USAR company personnel shall notify the IC and respond on the ground.

Generally, the first-due engine company will respond directly to the event and become the "landslide IC." The Landslide IC will establish the command post and work with law enforcement to clear the parking lot of vehicles to make room for a helispot and possible multi-casualty incident operation.

The first boat or helicopter on the scene will establish ocean IC. All other boats and helicopters shall report to the ocean IC for off-shore operations. All land-based companies shall report to the landslide IC. The personnel from these units will be assigned to assist with boat-based victim rescue/extraction, or landslide multi-casualty incident operations.

Swiftwater rescue

While en route, refer to the waterway rescue preplan for hazards and assignments.

- Request command-and-control to notify down-stream agencies.

- Be aware that your response may be a continuation of an upstream incident that already has an IC and an incident name.

- Be prepared to request multiple alarms to cover downstream sections of the waterway if the water is moving fast and the victim is likely to pass the downstream units on the first alarm. Be prepared to request enough alarms to cover the waterway(s) all the way to the ocean to give the victim the best chance of being intercepted.

Resource allocation

First alarm

- Five engines

- One EST: one truck, one squad, two helicopters, one USAR company, and one swift-water rescue team during storm deployment.

- One battalion chief

- *Additional alarms.* Request dispatch to send a first alarm using the location of the furthest downstream unit as the dispatch point. This will enable the dispatch of units in a linear pattern to cover downstream search/rescue points. Consider the potential for the victim to be swept into other channels downstream, and cover those also. The Helicopter may be assigned to search the entire waterway, and be prepared to attempt rescue if the victim is spotted.

Upon arrival

- Determine the point last seen (PLS) of the victim(s) and the time of sighting.

- Determine the number of victims, their clothing, ages, and other particulars.

- The first on-scene unit should place a marker float in the water for later downstream reference by other units.

- Name the incident and establish a communication plan. These incidents should generally be run on a mutual aid frequency for command because of multi-agency and/or multi-battalion operations.

Rescue operations.

- Appropriate safety equipment by all members operating near the water (no turnouts). Wear sneakers or running shoes instead of work boots and turnout boots. Other safety includes personal flotation devices, rescue helmets, etc.

- First responders should set up to conduct SAR using first responder techniques and equipment.

- Establish upstream safety and downstream safety positions

- Personnel assigned to USAR companies and swift-water rescue teams should be prepared to attempt rescue in the water using their special equipment and techniques.

- Shore-based or bridge-based rescue evolutions are the primary method of rescue because they do not require rescuers to enter the water and are often the most effective methods.

- High-risk rescue methods should be employed only after it is determined that shore-based and bridge-based rescues will not work and after a risk vs. gain evaluation determines that higher risk methods are required to accomplish the rescue.

Structure Collapse SAR

En route

Consider the possible cause of the collapse. All structure collapse operations have inherent dangers because of the instability of remnants of the building and debris piles. Collapse can result from fires, earthquakes, natural gas explosions, vehicles into structures, mudslides, floods, avalanches, or bombs. Each of these causes is often associated with particular hazards.

- If it is a major earthquake, you may be on your own, and there will be aftershocks.

- Always consider the potential for explosion(s) due to post-collapse gas leaks, secondary bombs (i.e., terrorist acts), etc.

- Consider any indication of terrorist bombing that might be accompanied by the release of nuclear, chemical, or biological agents.

- Consider precautions against the above-mentioned hazards.

- Review any pre-attack plans for the incident building(s).

Resource allocation

Collapse incidents are generally dispatched as *person trapped*:

- Two engines

- One truck company

- One squad

- One USAR company

- One hazmat task force

- One battalion chief

Upon arrival

Upon arrival to the incident, perform a size-up, noting the following:

- Size and occupancy of building and potential number of trapped/injured people.

- Construction of building and the type of collapse (pancake, lean-to floor collapse, lean-to cantilever, V-shape void, etc.).

- Presence of hazards (gas leaks, electrical, flammables, water main ruptures, bombs, etc.).

- Need for additional resources.

- Name the incident, request resources, establish command (be prepared to establish unified command).

Assignments

The following positions are options for *working* collapse incidents. Consult with USAR-1 personnel to determine which ones are needed for your incident:

- Safety officer (supported by lookouts stationed on aerial apparatus?)

- Rescue group leader

- Six-member void search squads

- Six-member rescue squads

- Search group leader

- Two-member search squads

- Add search dog teams to existing search squads as they arrive and integrate technical search capabilities

- Medical group leader

- Two-member medical teams for victims until they are extricated

- Multi-casualty incident (MCI) operations as necessary with the necessary MCI positions

- Technical group leader

- Heavy equipment/rigging specialist to advise on cranes and other heavy equipment and to act as liaison

- Shoring officer

- Cutting team

- Structures specialist (as necessary)

- Hazmat specialist

- Plans officer

- Logistics officer

- Equipment pool manager

In the case of a major collapse operation, the need for resources will rapidly outstrip the first alarm. The IC should not hesitate to request additional resources early in the incident if people are trapped and missing in a building collapse. Consult with the USAR company to help determine what resources are required.

Collapse assessment

Type of construction (wood frame, steel frame, tilt-up, etc.) and the type of collapse. There are four general categories of collapse types:

- *Lean-to floor collapse.* Occurs when one of the supporting walls fails or when floor joists break at one end. Usually creates large void spaces.

- *Cantilever.* Occurs when one end of the floor or roof section is still attached to portions of the wall. The other end hangs unsupported. This type of collapse is extremely dangerous because of unsupported sections.

- *V-shaped void.* Occurs when heavy loads cause the floor to collapse at center. Victims trapped above the V-collapse will usually be found in the bottom end of the collapse. Victims below the V-collapse will be found in the void spaces away from the V.

- *Pancake collapse.* Occurs when bearing walls or columns collapse, causing the upper floors to pancake down on the floors below. Victims may be found between floors or in voids created by furniture that is supporting the floors.

- When there is a total collapse, recognize the signs of impending secondary collapses:

  - Walls out of plumb

  - Smoke or water movement through bricks during fireground operations

  - Beams pulling away from walls

– Buckled or sagging steel beams

– Large cracks and plaster falling

• Abnormally low runoff, or soggy floors, during fireground operations

• Overloading or age

• Use standard USAR search markings to identify the results of your searches

See the attached diagram for search markings. Every building that has been searched should have the search marking spray-painted on the front of the building to signify that the search has been completed. Otherwise, other crews may waste valuable time searching the same buildings repeatedly. In a disaster situation, such redundant searching will drain valuable resources that could be used for other essential tasks. The result may be the loss of many lives that might have been saved elsewhere.

Five Stages of Collapse SAR

Stage 1: response, reconnaissance and surface rescue (first responder skills)

Conduct a general survey of the collapse area to determine the following:

• Building occupancy/use

• Number of potential occupants at the time of collapse and the number of victims reported trapped or missing

• What spontaneous operations are currently underway?

• Presence of hazards

• Structural stability of adjoining buildings

• Utilities shut off

Stage 2: immediate rescue of surface victims

Victims found on top of the debris or lightly buried should be removed first. Initial rescue efforts should be directed at removing victims who can be *seen* or *heard!* The next effort should be reaching victims whose locations are known even if they cannot be seen or heard.

Stage 3: void-space exploration, shoring, and rescue from likely places

Conduct an exploration of voids and other likely survival places:

• Search for trapped victims by looking in void spaces and other places that could have afforded a reasonable chance of survival when the collapse occurred

• Search for trapped victims using a *hail search*

• Place rescuers in *call* and *listening* positions

• Eliminate all external noises by calling for operational silence

• Going *around the clock*, each rescuer calls out or taps on something. A period of silence (and listening) should follow each call.

- After a sound has been picked up, at least one additional *fix* should be attempted to help triangulate the victim's position.

- Once communication with a victim has been established, it should be maintained until the victim is recovered.

Consider establishing void-space search squads,[1] a concept developed by some members of FDNY rescue companies in their efforts to streamline the process of void space SAR. In this concept, a search squad could consist of six persons, including a USAR-trained officer. The squad would be divided into two three-person teams—a *search team* and a *support team*. In this concept, the search team would perform the majority of the search operations, and the support team would provide material, shoring expertise, and other assistance as they burrow into the collapse area.

The search team. The void search squad officer is the supervisor and should generally be a company officer who is USAR-trained. He should assign the most experienced members to the search team. The officer triages the building to determine the most promising void spaces where live victims might be found. The most stable void spaces will allow maximum penetration with lower levels of risk to personnel.

If victim(s) are located, the officer will be responsible for determining appropriate extraction tactics, including calling in a rescue squad. It should be noted that the path created by the void search squad may not be the best path of egress for the victim. Removing larger sections of debris may open up better alternate egress routes.

The *void entry technician* is a position given to the first member to enter the collapse void. His main duty is to search the voids for the presence of victims and locate other voids within which victims may be trapped. Entering one void may lead to other, larger passageways within the collapsed structure. This is especially true where elevator shafts, hallways, and other access paths may still be partially intact.

As he makes his way into the void, the void entry technician removes debris and passes it out to waiting members. If possible, all debris should be removed from the void, not just moved to the side. This may require a bucket-brigade-type operation. Since the debris will likely have to be removed anyway, it makes sense only to handle it once.

The void entry technician may find it helpful to use small hand tools to cut rebar and pry debris away. Items such as search cameras may also be needed to look behind debris and into small void spaces for possible victims as the squad burrows into the collapse zone.

If a victim is found, the void entry technician should remove debris from the victim and make physical contact as soon as possible. A physical assessment should be made, with emergency treatment applied as appropriate (including treating for dehydration, shock, crush syndrome, inhalation of concrete dust, etc.).

The victim's identity should be obtained and transmitted back to incident command to establish victim tracking and possibly to help determine where additional live victims may be found (using housing or seating diagrams, etc.).

The following information should also be obtained and transmitted:

- The victim's location at the time of the collapse to help establish victim movement as the structure failed

- The type of victim injuries

- Whether other people were with or around the victim prior to the event and whether the victim heard other victims during entrapment and their possible location(s)

- Possibly, the cause of the collapse (i.e. explosion, etc.)

The *shoring technician* is the second void-space search squad member to enter the structure. He is responsible for assisting with debris removal and installing box cribbing or shoring to maintain stability of the collapse void. This position requires advanced skills in determining and building appropriate shoring as the squad moves further into the void space.

The support team. The support team consists of three members responsible for supplying physical assistance, tools, shoring materials, patient packaging equipment, etc. They are also the RIC for the search team. The *void-expander technician* assists the search team with ongoing operations; thus his tasks will vary according to the needs of the situation. He is primarily responsible for increasing the size of the path created by the search team. This, according to McConnell, enables additional personnel and equipment to enter the void space and provides better egress for victims after they are extracted.

The *support technician* goes by other names, including the *go-for* or the *relay person*. In essence, the support technician is responsible for gathering and moving tools and materials that are needed by the other squad members operating within the void space. He may also assist with expanding the void space, shoring, and other necessary support tasks.

The *tool and equipment technician* is responsible for establishing an equipment pool and a material pool to enable timely movement of equipment and materials as they are requested by other members of the void-space search squad. The tools must be tested for proper operation before they are passed to members working within the collapse zone. Broken and dirty tools must be rehabbed after use. In some cases, the tool and equipment technician must perform field repair of tools that are badly needed in the collapse. This member must maintain a tool log to ensure accountability of all tools during the operation.

Breaching and shoring. Breaching and shoring may be required to reach victims. The following are procedures and guidelines for breaching and shoring:

- Initially, try to avoid breaching walls because it may undermine the structural integrity of the rest of the building.

- It is safer to cut holes in floors and use the *vertical shaft* approach rather than to breach walls.

- If you must breach a wall or cut a floor, cut a small hole first to verify that you are not entering a hazardous area.

- Shoring may be used to support weakening walls or floors.

- Shoring should *not* be used to restore structural elements to their original positions. You just want to support them where they are now.

- Attempts to force beams or walls into place may cause collapse.

- Keep timber shoring as short as possible. The maximum length of a shore should be no more than 50 times its width.

- The strength of a shore depends on where it is anchored. If it's anchored to the floor, it depends on the strength of the floor to hold its load.

- Shoring should only be attempted by USAR personnel (or other qualified personnel) or under the supervision of USAR (or other qualified personnel).

- Shoring should *never* be removed once it is placed in an unstable building.

Stage 4: selected debris removal for SAR

Prior to this stage, all accessible void spaces have been checked and victims removed. Now it is time to reduce the size of the debris piles, searching for additional victims or to access victims we know are trapped below. Selected debris removal must be accomplished based on a *plan,* not haphazardly. Cranes and heavy equipment may be needed, under careful direction from USAR-trained heavy equipment/rigging specialists, to selectively remove heavy debris. Debris piles will be disassembled from the top down, in layers, looking for victims as the operation proceeds. Concentrate primarily on removing debris from areas where information suggests victims might be buried.

Stage 5: general debris removal

Stage 5 is essentially the demolition phase. General debris removal should be employed only after all other stages have been completed, or to remove material that's been searched to reach areas that require Stage 3 and 4 operations. The decision to begin Stage 5 should be made only after it has been determined that all victims have been accounted for, or that no other victims will be found alive. In many cases, Stage 4 has been reached only after 15 to 20 days.

Vehicle Extrication

En route

Based on the incident information (including the text), consider the need for additional resources. If the incident is reported to be a vehicle over-the-side, a semi-truck on top of a car, or a person trapped in a semi-truck accident, request a USAR company to augment the truck company. The USAR company can provide extra rescue air bags, cribbing and shoring for stabilization, special cutting tools, fiber-optic viewing equipment (to visualize entrapped extremities during extrication), and other special capabilities to expedite the rescue.

- If the incident is a vehicle beneath a semi-truck or other large vehicle, direct command and control to request the two closest heavy wreckers to support heavy lifting operations.

- If the incident is a vehicle into a lake or ocean, request a lifeguard division dive team, USAR-1, and a USAR truck.

- Consider the need for an air squad for rapid transport after extrication.

On scene

Follow standard techniques and methods, but remain flexible to adapt to changing situations. Vehicle extrication is an ever-changing topic partly because the design and engineering of modern automobiles is changing at a frenetic pace. The advent of driver-side air bags, head-protection air bags, the use of liquid natural gas and other new fuel and propulsion systems, and other innovations, create new challenges and hazards for firefighters and others who conduct vehicle extrication operations.

- Size-up for special hazards and needs.

- Determine extent of entrapment and develop rescue strategy.

- Stabilize vehicle (flatten tires, chock wheels, use cribbing and shoring to support overturned vehicles, remove fire hazards, etc.) to make the scene safe for the victim and rescuers.

- Provide safety hose line and traffic control.

- Begin treatment and extrication.

- For vehicles with air bags, take appropriate safety measures. Attempt to de-activate the air bag system. Before cutting or moving the steering column, wrap chain or a commercially-made safety device around the wheel to secure it from accidental deployment of the air bag. Consider the potential hazards from side air bags, head air bags, etc.

- Establish a shoring/cribbing team if the vehicle must be lifted or rolled up to free a victim. *Do not* lift or move vehicles without cribbing and shoring (lift an inch, crib an inch!) to prevent catastrophic movement of the vehicle.

- Assign a safety officer.

I Void-space search squads are a concept taught by Firefighter John O'Connell (FDNY Rescue Company 3) and Assistant Chief Mike McGroarty (California OES).

Glossary

basic operational level. The basic level represents the minimum capability to conduct safe and effective SAR operations at structure collapse incidents. Personnel at this level shall be competent at surface rescue, which involves minimal removal of debris and building contents to extricate easily accessible victims from non-collapsed structures.

basic rope rescue. Rescue operations of a non-complex nature employing the use of ropes and accessory equipment.

confined-space rescue. Rescue operations in an enclosed area, with limited access/egress, not designed for human occupancy and has the potential for physical, chemical, or atmospheric injury.

light operational level. The light level represents the minimum capability to conduct safe and effective SAR operations at structure collapse incidents involving the collapse or failure of light frame construction and basic rope rescue operations.

heavy floor construction. Structures of this type are built utilizing cast-in-place concrete construction consisting of flat slab panel, waffle or two-way concrete slab assemblies. Pre-tensioned or post-tensioned reinforcing steel rebar or cable systems are common components for structural integrity. The vertical structural supports include integrated concrete columns and concrete enclosed or steel frame that carries the load of all floor and roof assemblies. This type includes heavy timber construction that may use steel rods for reinforcing. Examples of this type of construction include offices, schools, apartments, hospitals, parking structures, and multi-purpose facilities. Common heights vary from single story to high-rise structures.

heavy wall construction. Materials used for construction are generally heavy and utilize an interdependent structural or monolithic system. These types of materials and their assemblies tend to make the structural system inherently rigid. This construction type is usually built without a skeletal structural frame. It uses a heavy wall support and assembly system to provide support for the floors and roof assemblies. Occupancies using tilt-up concrete construction are typically one to three stories in height and consist of multiple monolithic concrete wall panel assemblies. They also use an interdependent girder, column, and beam system for providing lateral wall support of floor and roof assemblies. Occupancies typically include commercial, mercantile, and industrial. Other examples of this type of construction type include reinforced and unreinforced masonry (URM) buildings typically of low-rise construction, one to six stories in height, of any type of occupancy.

heavy operational level. The heavy level represents the minimum capability to conduct safe and effective SAR operations at structure collapse incidents involving the collapse or failure of reinforced concrete or steel frame construction and confined-space rescue operations.

light frame construction. Materials used for construction are generally lightweight and provide a high degree of structural flexibility to applied forces such as earthquakes, hurricanes, tornadoes, etc. These structures are typically constructed with a skeletal structural frame system of wood or light-gauge steel components, which provide support to the floor or roof assemblies. Examples of this construction type are wood frame structures used for residential, multiple low-rise occupancies, and light commercial occupancies up to four stories in height. Light-gauge steel frame buildings include commercial business and light manufacturing occupancies and facilities.

medium operational level. The medium level represents the minimum capability to conduct safe and effective SAR operations at structure collapse incidents involving the collapse or failure of reinforced and URM, concrete tilt-up and heavy timber construction.

pre-cast construction. Structures of this type are built using modular pre-cast concrete components that include floors, walls, columns, and other sub-components that are field-connected upon placement on site. Individual concrete components use embedded steel reinforcing rods and welded wire mesh for structural integrity and may have either steel beam or column or concrete framing systems used for the overall structural assembly and building enclosure. These structures rely on single- or multi-point connections for floor and wall enclosure assembly and are a safety and operational concern during collapse operations. Examples of this type of construction include commercial, mercantile, office, and multi-use or multi-function structures including parking structures and large occupancy facilities.

search marking system. A standardized marking system employed during and after the search of a structure for potential victims.

state/national urban search and rescue (USAR) task force. A 62-person team specifically trained and equipped for large or complex USAR operations. The multi-disciplinary organization provides five functional elements that include command, search, rescue, medical, and technical. The USAR task force is designed to be used as a single resource and not disassembled to make use of individual task force elements.

structure/hazards marking system. A standardized marking system to identify structures in a specific area and any hazards found within or near the structure.

urban search and rescue (USAR) company. Any ground vehicle(s) providing a specified level of USAR operational capability, rescue equipment, and personnel.

urban search and rescue (USAR) crew. A predetermined number of individuals who are supervised, organized, and trained principally for a specified level of USAR operational capability. They respond with no equipment and are used to relieve or increase the number of USAR personnel at the incident.

References

California USAR Task Force 2, **After-Action Report on the Northridge Earthquake**, Los Angeles: County of Los Angeles Fire Department, March 1994.

Coleman, Ronnie J. **The Evolution of California Fire Service Training and Education**. Long Beach: California State University, 1994.

Company Officers Handbook, Los Angeles: County of Los Angeles Fire Department, 1997.

Downey, Raymond. **The Rescue Company**. Saddle Brook, New Jersey: PennWell Publishing Company, 1992.

Ellis, R. 1998. **Imagining Atlantis**, Alfred A. Knopf.

FEMA US&R Rescue Specialist Manual, Washington, D.C.: Federal Emergency Management Agency, 1998.

Gananopoulos, A. G. 1960. "Tsunamis Observed on the Coasts of Greece from Antiquity to the Present Time." **Ann. Geofis. 13:369-86**.

Lonsdale, Mark V. **Alpine Operations.** Los Angeles: Specialized Tactical Training Unit, 2000.

Sargent, C. 1999. "NFPA 1670: New Standards for Technical Rescue." **Fire Engineering,** (October 1999).

Index

C

D

M

N

O

P–Q

R

T

V

W–Y

Z

Shop Online & See Why People are Raving!

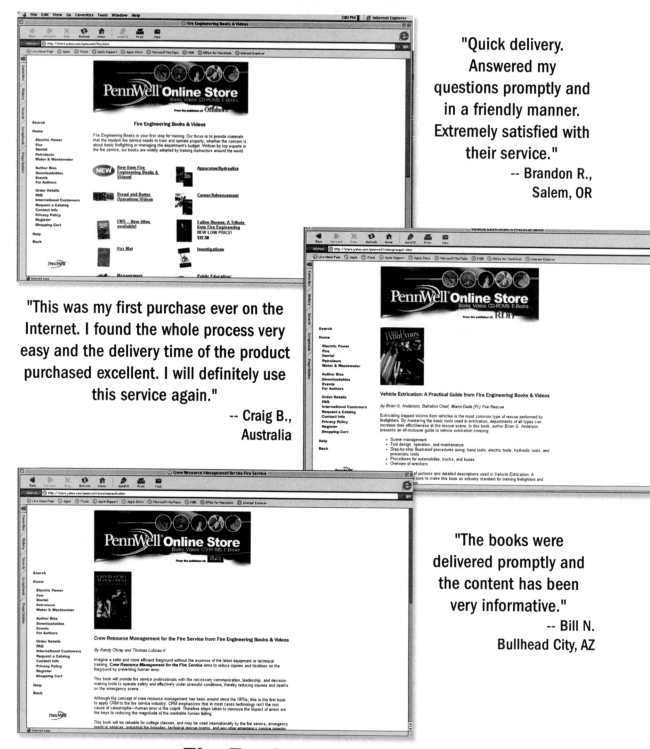

www.FireEngineeringBooks.com

3m